Green Homes and Workplaces

About the series Books for the Concerned Citizen

Leveraging the diverse expertise of its members in the subject domains and in publishing, the TERI Alumni Association proposes to publish a series of books on topics related to energy, resources, and the environment. The idea is to share information and, even more important, critical insights and understanding, with citizens who are keen to know more about some of the critical issues facing society and the world today but are lost in the deluge of information.

Our target audience is educated adults who are concerned about topical issues but lack the understanding to make sense of what they read or watch in the mass media—the series aims to equip them with conceptual tools and essential information not only to enrich their understanding but also to encourage them to act and thereby, albeit indirectly, further the UN Sustainable Development Goals.

The topics to be covered in the series and their respective subject-matter-specialist authors are listed below.

- **Rooftop solar**: Suneel Deambi and Shirish Garud
- **Coal**: Rakesh Kacker
- **Sustainable buildings**: Mili Majumdar and Minni Sastry
- **Electricity**: Sanjeev S Ahluwalia
- **Energy efficiency**: Ajay Mathur and Leher Thadani
- **Climate change**: Manish Shrivastav
- **Nutraceuticals**: Mayurika Goel
- **Public transport**: Shri Prakash and Sharif Qamar

The publication of this series is financially supported by the Shakti Sustainable Energy Foundation. The books will be published between February and October 2022.

Green Homes and Workplaces

Mili Majumdar
Minni Sastry

ISBN: 978-81-950776-4-9

Suggested citation
Majumdar M and Sastry M. 2022. *Green Homes and Workplaces.* New Delhi: TERI Alumni Association. 82 pp.

TERI Alumni Association
Administrative Wing, TERI
Darbari Seth Block (ground floor)
Habitat Place
Lodhi Road
New Delhi – 110 003

For more information, contact
Mili Majumdar (milimajumdar@gmail.com)
Minni Sastry (minnisastry@gmail.com)

CONTENTS

FOREWORD

Writing an article as a foreword to this magnificently brought out publication inspires me to terrains that I feel and think explores the human mind and action beyond time and space.

In great detail this publication investigates the role of humans on this Earth and their habitats. Over aeons of time as far as memory can stretch, the habitat has had the greatest impact on this planet and its ever-changing mode of life.

It is easy to argue that humans have erased the very nature that brought them up. But a little patience and thinking will reveal the contribution that our species have made. Yes, basically there are three forms of life: the plants, the animals, and the humans. It is evident to any searching mind that lives of plants are dependent on soil and care, accidentally or deliberately, in the space that gives them life or destroys them. As the Bible states, seeds are sown but the growth pattern is determined depending on the soil and care. Similarly the lives of birds and animals are dependent on the place of their origin and on their species. Only by accident can there be a change in their life form. They are a species and follow their life pattern within the specific terrain they are bound by. From the water of the seas and lakes to the terrain of soil and the air space they are disciplined by what they are. Yet it is great to experience that even in these defined spaces no two nests or ant hills are identical. They are same but different.

Now, the human species on the other hand can cross boundaries of time and space and in turn influence the very spaces. Yet as is evident by what one sees and learns, they with all the imagination and innovation within and behind them build identical repetitions. This is sad. It is time that architects and others who determine the built environment ensure that no two human living or work spaces are identical.

This is crucial, however much technology may dominate or art attempt to rule, the sheer composition of art and technology with the influence of time and space reflects the culture of the human

being differently and like a spectrum makes this Earth come alive.
Today one can build under water and anywhere on the soil as also
in space, and shortly on other planets and satellites. TERI is one
significant source of learning that makes life adventurous and worth
living. Our children learn from us but should not repeat us.

The challenge of change is fantastic. Climate is being
influenced by many factors. Not just our industrial and machine
life on the planet, but the very Earth is shifting its movement from
the Sun and the universe and in decades to come will see new
adventures that will change life form. Function will influence and
environment will make each and every one of us open new paths
and directions to ways of living habitat that we have never known
before.

Humans are born to explore, investigate and innovate with
imagination of infinity. Let us live.

June 2022

Krishna Rao Jaisim
Former chairman, Indian Institute of Architects,
Karnataka Chapter

'Little drops of water make a great, big ocean', and a clump of trees makes a forest—individual contributions, no matter how small, to any collective effort are of great value and significance. We all know that humanity faces existential challenges in the form of the changing climate, the ongoing pandemic, dwindling natural resources, and raging pollution. But the solution lies with us. We can make a difference and contribute to a better world. All of us can do our bit, and the collective impact of individual contributions is manyfold. However, when it comes to taking actions, we are often hampered by lack of information and, even more important, easy-to-use information sources that can tell us exactly what to do, how to do it, and what our actions would achieve. This booklet – a source of easy-to-use information – is meant to deal with that obstacle.

Do you realize that we spend 90% of our time inside buildings, be they our offices, homes, or places of entertainment and recreation? Are you aware that buildings are among the largest consumers of energy, water, materials, and land, and that about 40% of global emissions of greenhouse gases are attributed to buildings? Is it buildings that take up all these resources, or is it we who design, construct, and occupy buildings that fail to use resources efficiently and sustainably? Well, the responsibility lies with 'us', because it is for us, for our comfort and health and well-being, that energy, water, and other resources are expended. And most of that energy is generated from fossil fuels, and burning fossil fuels contributes to climate change and pollution. We also need water for bathing, brushing, cleaning and sanitation, but we use water recklessly. We buy food that is packed for our convenience in excessive and elaborate packaging—and we throw away not only that packaging but also food, just as we discard stuff that ends up in landfills. We prefer to drive, happy to be within our cocoons, instead of using public transport, and even for short trips we'd rather use cars and motorbikes instead of bicycles or walking.

The consequences of all these actions, small and large, add up, burdening the Mother Earth more and more.

Fortunately, all is not lost yet, and we have time enough to act wisely, save precious resources – energy, water, and materials – and if we do that, our collective impact will be humongous. Little things, such as switching to energy-efficient appliances, turning off taps when we do not need running water, using public transport or walking or cycling, reusing or re-purposing our belongings, and recycling waste to generate useful forms of energy, that we all can do can contribute immensely.

In this book, we explain some simple ways and means and offer simple tips as well as technical interventions to use resources more efficiently. To begin with, we give a glimpse of the problems cities face the world over and the actions some cities have taken to solve them. This overview is followed by detailed sections on energy-efficient design of spaces, concepts of thermal and visual comfort, and both passive and active means of keeping us comfortable. If our buildings are to use energy efficiently, we need climate-responsive designs, efficient appliances for air conditioning and lighting, and renewable forms of energy such as the sun and wind to generate power. We need clean air, ample sunlight, and thermally comfortable surroundings for healthy living. Many cities in India are grappling with severe air pollution, and we explain some key concepts related to air quality and suggest some simple ways to maintain good air quality. Water is a key resource that is growing increasingly scarce. Both quality and quantity of water are equally important, and the section on water focuses on the techniques to save water. We sincerely hope that you will implement these suggestions, and that we all will make the world a better place with our collective efforts.

Cities, Buildings, People

According to the United Nations, more than 50% of the world's population lives in cities at present and that by 2050, another 2.5 billion people will have joined it. About 90% of this increase will happen in cities in Asia and Africa. People migrate from villages to cities hoping for a better and healthier life. However, for how many do such dreams become a reality? Are cities geared to accommodate more people and meet their demands for resources? Let us take a tour of the city life, its challenges, and also see some simple and easy ways for citizens to contribute to making their cities and the built environment better places to live.

Countries of the developing world struggle to provide even basic amenities to their ever-growing populations. The amenities include housing, 24/7 electricity supply, transport, clean water, sanitation, and waste management. If the existing infrastructure of a city cannot cope with the demands, the city becomes increasingly polluted: the unmanageable municipal solid waste is dumped into landfills, wastewater gets mixed with natural streams, groundwater sources begin drying up, and air is fouled by smoke from coal burnt to generate electricity and exhaust gases from the ever-increasing number of cars and two-wheelers on the road.

Let us briefly overview the electricity consumption of a few sectors (Figure 1), the challenges they face, and the role citizens can play to make a difference and to contribute to sustainable living.

Energy

Energy is essential to development because it is the key resource for transport, agriculture, industry, buildings, water distribution, food production, etc.

If we look around, everything around us consumes energy: lights and fans in our homes, offices, and market places; cars, motorbikes, and buses we use for transport; motors that pump water into overhead tanks; the gas or electricity (or even coal and

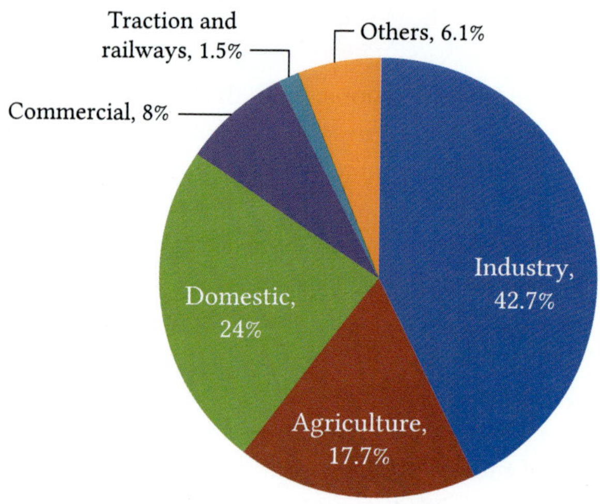

Figure **1** Consumption of electricity by sector: 2019/20

firewood) for cooking; and so on. Therefore, as global citizens and as end users we need to be careful, watch our consumption of energy, and use resources wisely. If we wish for a better environment, lower pollution, a booming economy, and assured supply of energy for our country, we need to conserve energy and use it more efficiently.

In this book, we focus on energy consumption in buildings – which accounts for approximately 32% of the total electricity consumed in India – and ways in which individuals can reduce that consumption in the workplace and at home and thus contribute indirectly to reducing pollution and the emissions of greenhouse gases.

Whereas in offices or commercial buildings lighting and air conditioning account for maximum consumption (Figure 2), in homes, it is comfort-cooling and lighting (Figure 3). Therefore, if only we design, build, and use buildings that use natural cooling and natural lighting, use energy-efficient appliances, and use renewable sources of energy such as the sun and wind, we can cut down our consumption of electricity substantially.

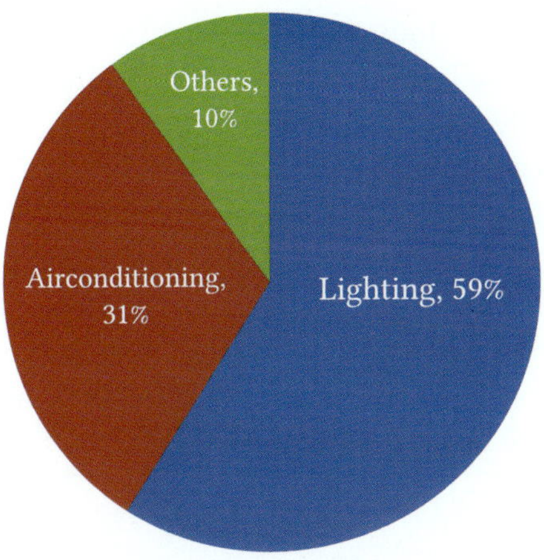

Figure **2** Electricity consumption in commercial sector, by end use

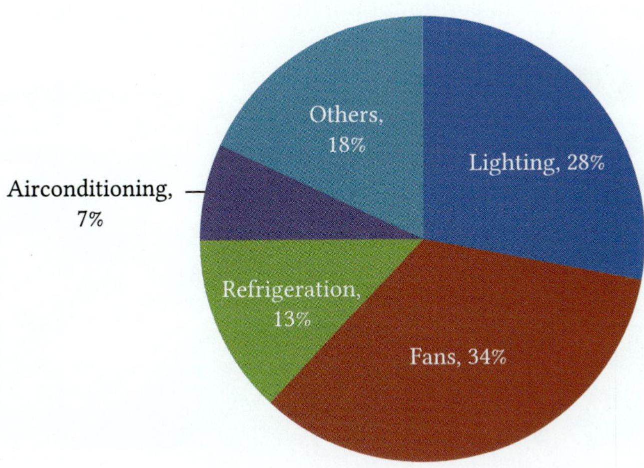

Figure **3** Electricity consumption in residential sector, by end use

An average home in India with an area of 100 square metres (about 1075 square feet) is estimated to consume about 5000 kWh (or billing units) of electricity a year. Assuming a community of 500 homes, this amounts to about 2.5 million units a year. It is also estimated that by using what is referred to as 'solar passive' design – orienting a building to minimize heating, using insulated walls and roofs, proper ventilation, etc. – and energy-efficient appliances in homes, this consumption can be easily halved, a saving that can provide electricity to 500 more such homes.

Here is a success story (Figure 4).

Colombia has been working to prepare its cities like Bogota, in which roughly 18 million Colombians are expected to move by 2050. The focus has been to make the city adaptive to climate change and low in consumption. Bogota has set a goal to improve

Figure 4 Bogota, Colombia (source https://www.rateitgreen.com/green-building-articles/the-7-most-sustainable-cities-in-the-world/102)

the quality of life in the city and to reduce the emissions of greenhouse gases from new buildings by 32%. Bogota has a public policy that flags buildings as a priority for reducing the emissions.

Transport

Transport is a sector that offers citizens the chance to make a difference by choosing more environment-friendly and sustainable mode, a choice that can help achieve healthy lifestyles, inclusive development, and diminished impact of climate change (Figure 5). According to the World Bank, the transport sector accounts for about 20% of the global emissions of greenhouse gases—if we continue to behave as though it is business as usual, the figure may reach 60% by 2050.

Emissions of carbon dioxide from privately owned vehicles can be reduced substantially by promoting public transport, making automobiles more efficient, and switching to electric

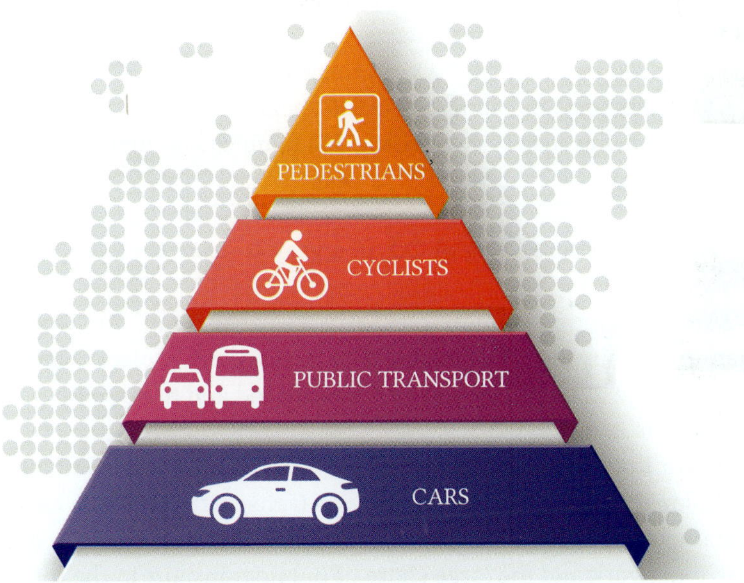

Figure **5** Sustainable transport pyramid

vehicles. **In India, transport is the third largest sector in terms of emissions of carbon dioxide; within the sector, road transport accounts for more than 90% of the emissions**. Responsible citizens who choose more sustainable modes of transport not only help the environment but also achieve the health benefits of walking or cycling. The closer the mode of transport to the top, the less its adverse impact on climate.

How personal modes of transport such as cars and motorbikes affect the environment

Assume the following figures for India.
• Average distance covered in one trip: 10 km
• Average fuel utilization: 12 kilometres per litre of petrol (source National Institute of Urban Affairs)
• Emissions of CO_2 per litre of petrol: 2.27 kg, which works out to 1.89 kg of CO_2 per trip.

If a family of six owns two vehicles, it will account for 3.78 kg of CO_2 emissions a day.

For 300 working days a year, this translates to 1135 kg of CO_2 emissions (and about Rs 50,000 spent on petrol).

A housing society comprising 500 families will therefore account for 570 tonnes of CO_2 a year.

Trees can absorb 21 kg of CO_2 in a year. (source https://www.viessmann.co.uk/)

Are we planting enough trees? Are we car-pooling? Are we planning to buy electric vehicles? Our actions and choices could make a huge difference to our neighbourhood and to our city.

Here is a success story.

Hong Kong was ranked first in the Sustainability Cities Mobility Index in 2017, making Hong Kong's transport system the most sustainable globally. This success was due to the well-connected metro network and the large share of public transport in the total number of trips. **Public transport in Hong Kong is cheap: less than one-fifth of people in Hong Kong own a car** (Figure 6). (source https://hongkongbusiness.hk/transport-logistics/news/hong-kong-has-most-sustainable-transport-system-globally)

Figure **6** High share of public transport in Hong Kong

Solid waste

In 2016, cities worldwide generated about 2 billion tonnes of solid waste, which means each city dweller generated approximately 750 grams of solid waste every day—and the total is expected to increase by 70% by 2050 (source the World Bank). Developing countries and low-income countries face the challenge of effectively managing this mountain of solid waste. About 90% of municipal solid waste is either dumped in landfills or on open land or burnt. Untreated solid waste poses several challenges to the city administration in keeping the city clean and odour-free, removing hazardous waste, and protecting the environment from methane, a greenhouse gas, which is generated by untreated wet and organic waste.

Citizens can help a great deal to reduce the burden of waste on Mother Earth by reducing the amount of waste we generate in the first place, segregating the waste we generate, managing its treatment and disposal, and turning waste into resources (Figure 7). These ways include minimizing the use of plastic, converting kitchen waste and other organic or wet waste into compost at every level from individual homes to businesses to entire cities, avoiding wastage of paper, recycling paper and other recyclable waste, and segregating waste into its recyclable, hazardous, electronic, and organic components, because each such category can then be managed more efficiently.

Figure 7 Segregating waste at the level of individual homes into wet, dry, and recyclable waste

Some cities around the world are inspiring examples of how waste can be reduced and managed through innovative policies designed by the governments and implemented successfully by the citizens. For example, San Francisco recycles or composts more than 80% of the waste it generates and thus diverts most of the waste from ending up in landfills. Some of the measures adopted by the city include making businesses and residents segregate recyclables and food waste, banning polythene bags in retail and grocery shops, and recycling construction debris. San Francisco's vision of a zero-waste future has won the support of its residents and businesses, all of whom are united in keeping garbage out of landfills.

A family of four generates about a tonne of waste every year. Bengaluru, with a population of 12.3 million, would generate 123,400 tonnes of waste every day. In volume, that waste is equivalent to that of a building 10 storeys high and spread over 4000 square metres (roughly an acre).

Water

As the population of a city increases, it consumes increasing quantities of fresh water and generates increasing quantities of wastewater. By 2050, water demand in cities is estimated to increase by 50%–70%. Water scarcity in developing nations threatens social stability, has even led to riots, and is a challenge for farmers and particularly the poorer city dwellers. Converting water scarcity into water security is the key to healthy, peaceful, and sustainable living.

As for water management, some cities have shown the way. Morocco, for example, encourages the management of groundwater through participation and plugging the leaks in its water supply network. Orange County, in California, has reduced its dependence on external sources of water by reusing treated wastewater. Other measures include recharging groundwater through harvesting rain water, encouraging businesses and housing societies to install wastewater treatment plants of their own, and building the requisite infrastructure to reuse treated water.

While water becomes increasingly scarce as years go by, we are also likely to demand more and more of it as standards of living rise—a situation that is clearly unsustainable. Let us see how, as responsible citizens, we can reverse the trend and strive to make our country water secure.

Conventionally, water demand, or the amount of water each one of us needs daily, is assumed to be 150 litres on average. However, if we decide to use water more efficiently, that figure can be brought down by about 50% to only 73 litres—the table shows how.

In a residence, with conventional inefficient water fixtures, consumption per person for different uses is as follows.

Site or standard fixture	Flow rate (litres per minute)	No. of uses	Consumption (litres)
Toilet	(per flush) 9	8	72
Lavatory faucets	4	10 × 0.5	20
Shower	10	1 × 5	50
Kitchen faucets	4	3 × 0.5	6
Total water consumption	—	—	148

By switching to more efficient fixtures and using separate fixtures for the disposal of solid waste and liquid waste, the consumption can be cut by 40%.

Site or standard fixture	Flow rate (litres per minute)	No. of uses	Consumption (litres)
Toilet	(per flush) 6	2	12
Water closet (liquid only)	(per flush) 3	6	18
Lavatory faucets	2	10 × 0.5	10
Shower	6	1 × 5	30
Kitchen faucets	2	3 × 0.5	3
Total water consumption	—	—	73

Climate, Built Environment, and Comfort

Hotter summers, cooler winters, and more frequent extreme weather events are all manifestations of global climate change. Increasingly, the trend is to design buildings without considering the climate but to modify the microclimate instead through space heating and cooling—a trend that leads to increasing demand for electricity. For example, in the composite climate zone, with hot and dry summers and severe winters, the modern glass and steel buildings tend to radiate heat inside of buildings in summers, making the microclimate uncomfortable without air conditioning whereas in winters, windows let in the cold air from outside into the buildings, which therefore require rooms to be heated. Thus, buildings are increasingly dependent on air conditioning to keep the indoor temperatures within comfortable limits. Air conditioning in buildings accounts for about half their total energy consumption. This way of achieving indoor comfort is both expensive and energy intensive.

However, buildings can also be designed to be energy efficient, providing comfortable living and working spaces while minimizing energy consumption and the negative impact on the environment of excess consumption.

Given India's predominantly tropical climate, the requirement for cooling are far more than those for heating. We feel comfortable so long as we can dissipate heat from our body into the surroundings easily—and uncomfortable when we cannot.

Buildings should be designed to create best possible indoor conditions so that those live or work within the buildings feel relaxed and comfortable and remain healthy.

Comfort inside buildings is defined by several parameters, the more important being air temperature, relative humidity, air movement, light level, noise level, and the quality of indoor air. Indoor conditions and comfort are highly dependent upon the outdoor climate and microclimate.

Efficiently designed buildings respond to the outdoor climate with integrated design features that ensure indoor comfort with minimum consumption of resources.

Thermal comfort

Thermal comfort expresses the state of mind and perception of people, whether they feel cold or hot or comfortable in an environment. The perceived comfort is greatly influenced by such external factors as air temperature, humidity, radiation and wind speed as well as such personal factors as clothing and physical activity. The external factors, in turn, are influenced by the climate, the location, the built environment, building design, and even the overall plan (or lack of it) of the city itself. Well-designed cities with green infrastructure, properly designed open spaces, optimized urban density, adequate transport, and buildings designed in harmony with nature give us spaces that are thermally comfortable.

Thermal comfort inside buildings is greatly influenced by the building's architectural design. For example, a home in Delhi with windows that face the north and the south is likely to be more comfortable in summer as well as in winter. However, if the same house were to have large unshaded and fixed windows facing south-west and west, it is likely to be particularly uncomfortable in summer months because of overheating of its internal spaces. **More than 50% of electricity used in buildings is for achieving comfort,** either in terms of temperature (for cooling or heating) or in terms of light (ambient lighting, spot lighting for close work, night light, etc.). Comfortable indoor spaces also make people more productive, energetic, alert, and healthy.

Parameters of thermal comfort

Of the many factors that determine how comfortable the occupants of a building are, the most important are air temperature, air speed, relative humidity, mean radiant temperature, clothing, and the body's metabolic rate. Each of these parameters has a distinct unit by which the parameter is

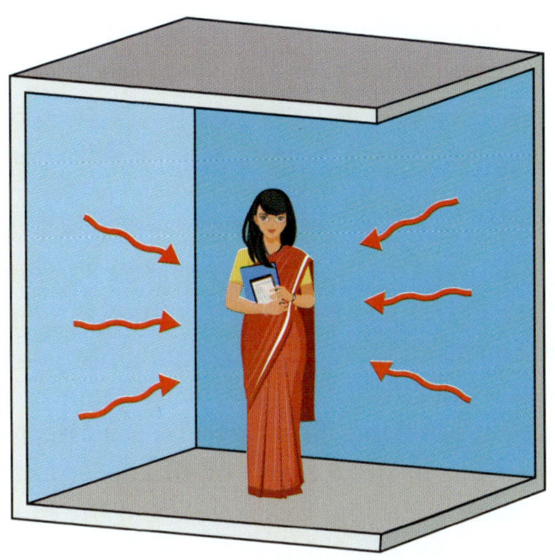

Figure **8** Heat transfer through radiation between building
envelope and occupants

measured; temperature, for example, is expressed in degrees
Celsius; relative humidity is a percentage, and air speed can be
given in metres per second or in kilometres per hour. Metabolic
rate is measured in met value, which varies depending on activity:
sleeping, for example, has a met value of 0.7, equal to a heat
output of 76 watts; walking has a met value of 2.0, equal to 212
watts; and light to moderate exercise can have a met value of 3.5,
equal to 370 watts. Clothing is measured in clo units: light clothes,
for example, have a clo value of 0.30, whereas garments that cover
the whole body have a clo value of 1. Clothing can have a huge
impact on thermal comfort: in summer and working in an office,
wearing a long-sleeved shirt and a tie can make you feel very hot,
whereas a light cotton T-shirt will be far more comfortable.

While clothing and metabolic rate are factors you can
control, air temperature, relative humidity, and air speed are
external factors over which you have little control. That is where
good design makes a difference by using techniques – usually
referred to as 'passive solar' – that incur no running costs

(unlike fans and air conditioners) and yet serve to moderate air temperature, relative humidity, and air speed to ensure a degree of thermal comfort.

Another key component that impacts our sensation of comfort is mean radiant temperature. The level of thermal comfort we experience indoors is a function of air temperature and the temperature of the surfaces in that space (Figure 8), represented by the mean radiant temperature, which, in turn, is a function of solar radiation at a given location, the design of the building, and building materials—all of which determine the temperature of internal surfaces in a building. **Solar passive architecture, by using such methods or devices as shading, orientation, mass of materials (thick walls, for example), and the materials themselves, can make a big difference to the temperature of external surfaces, thereby changing the temperature of internal surfaces as well and thus contributes to thermal comfort**—by minimizing heat gain in summer and heat loss in winter, for example to ensure thermal comfort without using any appliances. Another technique of solar passive architecture is evaporative cooling, which is achieved through water bodies such as pools and fountains, plants including shade trees, and more effective natural ventilation by means of appropriately designed openings, and can contribute greatly to modifying the outdoor environment for our comfort. On the other hand, buildings designed without regard to the climate of a city and the microclimate of the neighbourhood depend a great deal on external appliances for thermal comfort.

If the internal surface temperature is higher than that of the human body (36–37 °C), occupants end up gaining the heat being radiated from those surfaces and feel the discomfort. Similarly, if the temperature of the ceiling and of the walls is lower than the body temperature, we feel cold. Solar passive architecture revolves around this fundamental principle.

The concept of mean radiant temperature helps us to understand urban heat islands. Have you noticed that cities full of buildings are hotter than the countryside, which is full of open spaces, shrubs, and trees? This is simply because the hard surfaces of buildings absorb heat and radiate it into their surroundings, making the ambient air warmer.

Here is a demonstration of the urban heat island effect in Bengaluru (Figure 9). The maximum difference in air temperature between two locations at the same time of the day was 7 °C. The coolest location, the campus of the Indian Institute of Science, was a site dense with trees and with buildings that were designed to be climate responsive design, whereas the hottest location, the Electronic City, was a site with sparse vegetation and modern buildings with large glass walls and hard concrete surfaces—a perfect heat island. Comfortable outdoors, on the other hand, usually mean comfortable indoors.

To summarize, designing buildings based on climate contributes to thermal comfort with minimum dependence on electricity, which not only makes such buildings more environment friendly but also cuts down running costs.

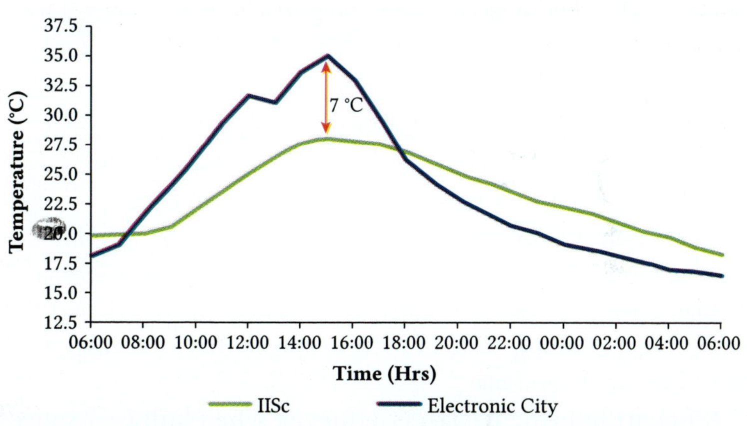

Figure **9** Difference in hourly air temperature at different locations in Bengaluru that differed in the amount of vegetation

The design and materials used for the 'building envelope' (which refers to the outer walls, windows, and the roof) are crucial to heat gain or loss, depending upon the climate. The building envelope moderates the impact of the outside weather similar to a bowl of hot soup that quickly turns lukewarm and then cold by losing its heat to the surroundings or a bottle of cold water, once taken out of the refrigerator on a hot summer day, becomes warm—if neither is thermally insulated (Figure 10).

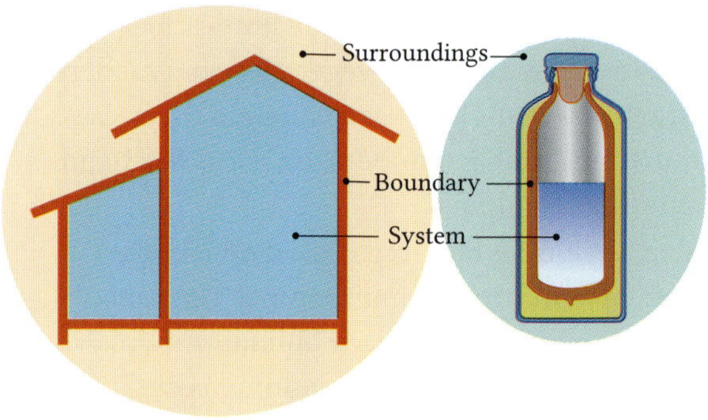

Figure **10** A building envelope compared to an insulated flask

Visual comfort

Light is the visible form of electromagnetic radiation. Different spaces within a building or a home require different levels of light depending on what they are used for. For example, a kitchen requires a different level of lighting than what a living room requires. The lighting requirements for different spaces are standardized by technical groups. These requirements are part of the National Building Code of India to ensure that all buildings meet those requirements.

Efficient lighting design is achieved when building spaces are optimally lit with minimum consumption of electricity. One way to reduce dependence on electricity is to make more use

As temperatures and humidity rise, one of the ways to feel comfortable is to make air flow faster around human body so that the heat being dissipated is carried away quickly. However, a stronger current of air may also be a nuisance; for instance, it can send loose papers and other light and unanchored objects flying or make the surroundings noisy. The recommended 'speed limit', therefore, is 2 metres per second (or 7.2 kilometres per hour, which is really quite low).

For example, the National Building Code stipulates that if air temperature is 30 °C and relative humidity is 60%, air speeds up to 0.1 m/s are good enough for thermal comfort. However, to remain within the comfort range, at the same level of humidity, the air speed needs to go up to 1.6 m/s if the temperature goes up to 32 °C. On the other hand, if the temperature remains at 30 °C but relative humidity goes up to 80%–90%, air speed up to 0.6 m/s is good enough (Table 1).

Table **1** Desirable wind speeds (metres per second) at different combinations of air temperature (29–35 °C) and relative humidity (30%–90%) to achieve thermal comfort

DBT[1]	Relative humidity (%)						
(°C)	30	40	50	60	70	80	90
29						0.06	0.19
30				0.06	0.24	0.53	0.85
31		0.06	0.24	0.53	1.04	1.47	2.10
32	0.2	0.46	0.94	1.59	2.26	3.04	
33	0.77	1.36	2.12	3.00			
34	1.85	2.72					
35	3.2						

[1] dry bulb temperature   source *National Building Code of India*

of natural daylight (Figure 11), which is also a source of light of good quality. India is blessed with ample sunshine, which should be fully used in buildings to meet lighting requirements during daytime. Nowadays, people spend most of their time indoors, which makes proper lighting, including daylight, particularly important for the good health of occupants.

However, natural daylighting does not include direct sunlight, because it creates such problems as shadows, overheating of

Figure **11** Diffuse daylight integration in a shopping mall

enclosed spaces, and uneven distribution of light. **Natural daylight is diffuse light, free of glare, which is achieved by using external shading structures that also reduce heat gains, thereby contributing to thermal comfort as well**.

Natural daylight to achieve visual comfort
The source of natural daylight is the sun. Sunlight reaches the insides of a building through many ways. However, the most useful form of daylight is diffused light from the sun, the sky, and the surroundings during daytime. Let us look at some ways of designing our buildings to help integrate natural daylight into the lighting system (Figure 12).

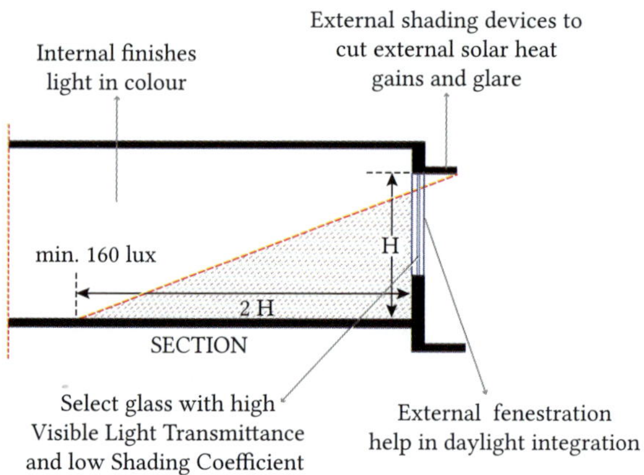

Figure **12** Integrating daylight into building design

Designing artificial lighting

In a sustainable or 'green' building, artificial lighting is provided after daylight hours or when daylight levels are not adequate to carry out the required tasks. **'Green' lighting design is meeting the lighting levels for a space with the most efficient artificial lighting design, selection of most efficient luminaires, and keeping in mind human-centric lighting that facilitates the circadian rhythm and thus health of occupants.**

Illumination levels for different tasks are recommended by such codes and standards as the National Building Code of India. It is important that spaces are neither over illuminated nor under illuminated. The level of illumination at a given space, or how much light falls on a particular surface, is defined in terms of lux. Think of 1 lux as the level of illumination you can expect over an area of 1 m² if it is lit by a single candle placed 1 metre away from that surface: in other words, lux is determined not only by (1) how bright the source of light is but also by (2) how far away it is from the surface to be illuminated and (3) how large that surface is. For example, for general office space, the minimum recommended

level is 300 lux, a middle level is 500 lux, and the maximum is
750 lux. Similarly, for homes, the general lighting recommended is
50–150 lux. On a full moon night and a clear sky, the level is 1 lux.

Whereas light levels are aimed at visual comfort, lighting
efficiency takes into account the amount of energy consumed and
is expressed as LPD (lighting power density) in the building code.
That density at a given space is obtained by dividing the total
connected load for lighting (in watts) by the floor area (in square
metres). **'Green' lighting design strives to achieve the desired
level of visual comfort at the lowest LPD by using a variety
of energy-efficient lamps and luminaires in the market**.

However, as explained earlier, besides the source of light,
room dimensions as well as the finish and colour of surfaces,
or their reflectance, are important in achieving efficient visual
comfort. Lighter colours for walls, ceiling, and the floor are
preferred because they reflect more light.

Lighting for well-being

The circadian rhythm is our body's internal clock that tells us
when it is naturally the right time to turn in at night or to wake
up in the morning. The body clock helps in regulating sleep
patterns, feeding behaviour, release of hormones, blood pressure,
and so on. Light – more specifically, daylight – affects the body
clock and thus the circadian rhythm significantly. Put simply, the
circadian system is a blue-sky detector. Exposure to lots of blue
light tells the body to be awake and alert. The dominant colour of
the sky and daytime sunlight is blue, and the shorter-wavelength
blue light has a greater impact on our circadian system than the
long-wavelength red or warm light has. We need cool white light
with a high blue content and brighter illumination to remain more
alert and active during the day (Figure 13), and we need warm
white light with greater red content in the evening to enjoy good-
quality sleep (Figure 14). We wake up with daylight and sleep with
sundown. Thus light plays a larger role in our daily lives beyond
enabling us to see.

Figure **13** Optimal electrical lighting combined with daylighting
to keep students alert and active (a school rated
Platinum, the highest rating, by the US Green Building
Council)

Human-centric lighting, or lighting that is in line with
the circadian rhythm, goes beyond illumination levels, lighting
efficiency, and uniform distribution. Lighting parameters affect
circadian rhythm, which are reflected in the levels of melatonin
– its production or suppression – and are different from those
that impact visibility. Whereas rods and cones are photoreceptors
that help us to detect colour, brightness level, and movement,
yet another set of photoreceptors, namely IPRGCs (intrinsically
photosensitive retinal ganglion cells) is critical to maintaining the
circadian rhythm (Figure 15).

These photoreceptors are sensitive to teal-blue light (480 nm)
and measure the amount of blue light to help our body to sense

Figure **14** Warm light in homes in evenings facilitates sleep

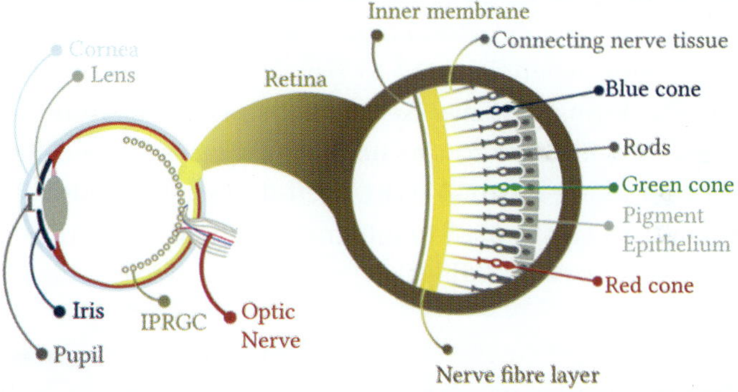

Figure **15** The third photoreceptors – IPRGCs (intrinsically photosensitive retinal ganglion cells) – critical to the maintenance of the circadian rhythm

the time of day. They are connected to SCN (suprachiasmatic nucleus), which is the part of our brain that houses the body's master clock and works in collaboration with the pineal gland and other hormones to regulate the circadian rhythm and hormone cycles and, in turn, regulates many physiological aspects such as alertness, digestion, and sleep. Overexposure to blue light in the evening disrupts the production of melatonin and thereby disrupts the sleep–wake cycle, with direct adverse impacts on health and productivity at work.

A dynamic lighting system works in tune with the circadian rhythm. Such a system is a blend of the right kind of light and desirable spectral distribution, ensures proper distribution of light in horizontal and vertical planes, and is equipped with intelligent controls. Daylight is a greatly desirable component of such a system. For further details including the threshold levels of the various parameters of visual comfort, refer to ISHRAE's Indoor Environmental Quality Standard. (ISHRAE is short for Indian Society of Heating, Refrigerating and Air Conditioning Engineers.)

Switch your mobile phone to a warmer mode in the evenings. It may be the culprit that does not let you fall asleep.

Integrating daylighting with artificial lighting
Daylighting is letting in natural light from the sun and the sky to meet the requirements related to visual comfort and lighting requirements during daytime. **According to the National Building Code, at least 25% of occupied spaces in a building should be lit using natural light**. Well-designed windows and lighting controls can be used to integrate daylight with other sources of light inside buildings so that the energy consumed on lighting is reduced.

Tips for building owners
Clear goals need to be set to use daylight as part of the design. Spaces that require more daylight should be placed at the periphery of buildings. Such spaces include workstations in an

office building and the kitchen and the study room in homes. Also, because patients benefit from natural daylight and outside views, health-care facilities should use daylighting.

Spaces such as auditoria and theatres or cinema halls in which lighting needs to be closely controlled – and therefore do not benefit from daylight – can be placed closer to the centre of a building.

Narrow buildings benefit more from daylight than broader buildings do, because daylight penetrates deeper into narrow buildings. Typically, useful daylight let in by a window serves a space up to twice the height of the window in terms of the distance from window. In other words, incoming daylight from a 1 metre tall window will illuminate the space within up to 2 metres away from the window.

Glare-free daylight is meaningful daylight, and it is possible to ensure that through external shading devices. If such devices are missing, direct sunlight that penetrates living spaces will be so bright that it causes glare. To avoid glare, the occupants typically pull down the blinds or internal louvers and switch on the lights, defeating the very purpose of windows and increasing electricity consumption for artificial lighting.

Air quality index

Air quality is of significant concern in Indian cities, and their depleting air quality and its impact on health are issues widely covered in the media. Air quality is a complex measure, which is why it is expressed using an index, namely AQI, or the air quality index.

What is AQI (air quality index)?

Air quality index is a number to measure and communicate the level of air pollution: **the higher the AQI, the worse the impact of pollution on human health**. The index is based on the concentration in a given volume of air of eight pollutants at a given location. The pollutants are as follows: particulate matter with particles 10 micrometres (µm) or smaller or 2.5 µm or smaller (PM_{10} and $PM_{2.5}$, respectively), oxides of nitrogen (NO_x), ozone (O_3), carbon monoxide (CO), sulfur oxide (SO), ammonia (NH_3), and lead (Pb). (A micrometre (μm), also known as a micron, is one-thousandth of a millimetre.) Based on the measured concentrations of these pollutants in ambient air, a sub-index is calculated for each of the eight pollutants. The average concentration over 24 hours is taken for all pollutants except CO and O_3; for these two, it is the average of 8 hours. All eight pollutants may not be monitored at every location (the overall AQI is calculated only if data are available for at least three pollutants, one of which must be $PM_{2.5}$ or PM_{10}).

The key point to remember is that even if the concentration of one critical pollutant is high to severe, the overall AQI will be high, irrespective of the concentrations of the other pollutants.

Air quality index can range from 0 to 500 or even higher. A value of 0–50 denotes fresh air, good for health, and that above 400 signals very bad air, prolonged exposure to which can lead to premature death. New Delhi recorded a minimum AQI of 271 and a maximum AQI of 674 in 2021—people in Delhi breath air that ranges from unhealthy to hazardous all the year round.

Health impact of air pollutants

Air can be polluted with many substances of varying sizes and chemical composition.

Coarse particles greater than 10 μm in diameter are generally trapped in the nose itself – they are too large to travel further along the respiratory tract – but can cause diseases of the nasal and upper respiratory tracts.

Fine particles, smaller than 2.5 μm in diameter, penetrate deeper, reaching the lungs, and may cause heart attacks, strokes, asthma, and bronchitis and can even impact early brain development in children.

Black carbon or soot, a particular kind of particulate matter that comes from burning fuels (especially oil, wood, and coal), has serious health impacts, leading to heart attack, hypertension, asthma, COPD (chronic obstructive pulmonary disease), bronchitis, and various forms of cancer.

Oxides of nitrogen (NO_x) are a combination of NO (nitrogen oxide) and nitrogen dioxide (NO_2) formed at high concentrations around roads (because of exhausts from vehicles) and can cause asthma, bronchitis, and heart diseases.

Ozone at ground level is formed through reactions between VOC (volatile organic compounds) and nitrogen oxides; ozone is a respiratory irritant and can cause chest diseases.

Sulfur dioxide is emitted by burning fossil fuels that contain sulfur and causes irritation of the eye, respiratory diseases, and cardiovascular diseases.

Outdoor air enters indoor spaces through open windows and doors. In regions in which the quality of outdoor air is poor, ensuring that indoor air is healthy is an even tougher challenge because the outdoor air itself pollutes indoor air and, once it enters confined spaces, the concentration of pollutants can be higher indoors than what it was outdoors. Outdoor air can bring in particulate matter and emissions from vehicles. In addition to the pollutants thus brought in, indoor spaces have their own sources of pollutants such as VOCs given off by building and furnishing materials, chemicals from cleaning agents and room fresheners, tobacco smoke, indoor combustion, printers and copiers, kitchen fumes, and biological contaminants such as moulds, mildews, bacteria, and house dust mites. Viruses such as Sar Cov 2 are also airborne and, given the COVID-19 pandemic, need special attention.

An adult typically breathes in 12,000 litres of air every day, an input vital for survival, let alone health.

Controlling air quality

The quality of indoor air can be improved through proper ventilation and by filtering the incoming air. However, controlling indoor sources of pollution is equally important. All potential sources of VOC should be controlled, and the concentrations of hazardous VOC and SVOC (semi VOC), HFRs (halogenated flame retardants), urea-formaldehyde, and select phthalates commonly used in building materials and products should be within acceptable limits. For example, HFRs in building products should be less than 100 ppm (parts per million) or below the maximum permissible limit stipulated by the local code. Common sources of HFRs are furniture, flooring, ceiling tiles and wall coverings, piping and electrical cables, conduits and junction boxes, etc. Similar restrictions should be enforced in the case of urea-formaldehyde found in composite wood products, laminating adhesives and resins, and thermal insulation. Phthalates, found in vinyl flooring and consumer products such as shampoos, soaps, and hair sprays, are equally harmful and their use should be restricted. Declaring the presence of harmful compounds, if any, in materials is a commonly adopted norm in several counties, and it is high time that we pay equal attention to all these contributing factors to indoor air pollution—after all, for many, nearly 90% of the life is spent indoors.

Among the strategies to improve the quality of indoor air are

o controlling the pollutants at source
o managing the sources of pollution
o customizing ventilation and filtering
o maintaining and operating the above measures effectively.

The quality of indoor air can be improved in many ways.
Monitor the quality of outdoor air The quality of local outdoor air at the site determines, for example, the approach to ventilation, namely mechanical, natural, or a mix of approaches. In cities such as Delhi in which the quality of outdoor air is always poor,

mechanical ventilation with extensive filtering is essential. If the concentration of PM_{10} is more than 50 µm per cubic metre, such air should be used with caution.

Select what pollutants to filter out Finding out sources of contamination is another critical aspect of ensuring good air quality. Buildings close to dense traffic, for example, need to pay more attention to NOx, CO, and ozone besides particulate matter and VOC. The concentration of particulate matter and the distribution of particles of varying fineness are key to choosing suitable filtration systems. For example, if the outside air contains more of coarse particles, high-efficiency filters such as MERV 13, which keep out particles larger than 1.0 µm (Figure 16) used alone will not work well because such filters will quickly become clogged: filtering the air at several levels will give better results; such a system will involve pre-filtration to remove coarse particles, followed by filtering the pre-filtered air through higher efficiency filters. Activated carbon filters are also recommended to remove VOCs.

Design for positive pressure Maintaining a positive pressure in mechanically ventilated spaces can keep pollutants out whereas

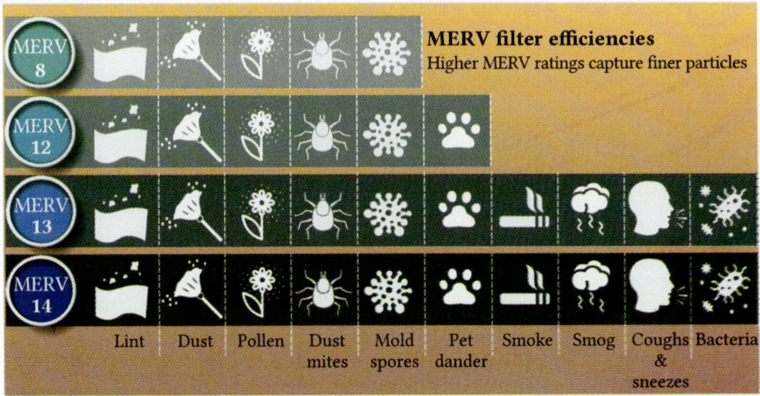

Figure **16** Rating of MERV (minimum efficiency reporting value) filters; the higher the rating, the finer the particles trapped by the filter.

negative pressure in a building allows infiltration through openings, crevices, and joints. The volume of incoming outdoor air should be at least 10% greater (or as recommended by the consultant engaged for mechanical, electrical, and plumbing-related matters) than the volume of outgoing air leaving a building through its exhaust outlets so that a positive pressure is maintained (Figure 17).

Use green plants Some common indoor plants, especially those grown for their decorative foliage, such as the rubber plant (*Ficus elastica*) and the areca palm (*Areca catechu*), offer a natural way of removing such toxic pollutants as benzene, formaldehyde, and trichloroethylene from air, helping to combat the sick building syndrome. However, indoor plants cannot keep out fine particulate matter ($PM_{2.5}$ and PM_{10}): only appropriate filters can keep it out.

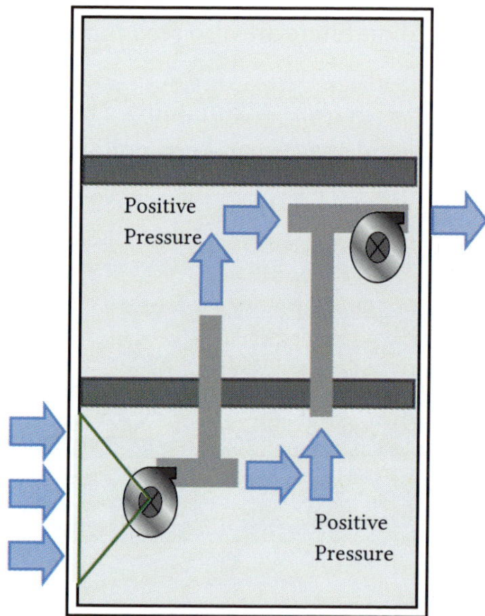

Figure **17** Design for positive pressure (which is possible only if the volume of air coming in is greater than that of air going out or being distributed within a building)

Water

Water is a dwindling resource, and millions of people in India live without access to water even to meet their daily needs. Hence it is everybody's responsibility to use water judiciously and avoid wasting this precious resource.

Measures to save water

Be conscious of how much water you use for your daily chores. Spend less time under a shower. Instruct household help not to keep taps running while washing clothes, dishes, and pots and pans.

Fix leaky taps or joints, if any, immediately.

Water saving aerators (Figure 18) can save up to 80% water while giving the same level of cleaning every time

- You waste 5 litres of water every time you brush your teeth, barely awake and groggy, and leave the tap on. For a family of 4, it amounts to 20 litres of water wasted every day merely for brushing.
- Showering for even 10 minutes can consume up to 120 litres of water.
- Washing clothes under a running tap requires up to 20 litres per person.
- Doing the dishes and washing pots and pans under a running tap twice a day can take up to 50 litres of water per person.
- Flushing uses up another 45 litres.
- Cooking needs about 5 litres per person.
- Other uses, such as gardening, washing a car, and household cleaning, consume even more water.
- The recommended upper limit in India is 150–200 litres a day per person in cities (assuming the use of a flushing toilet).

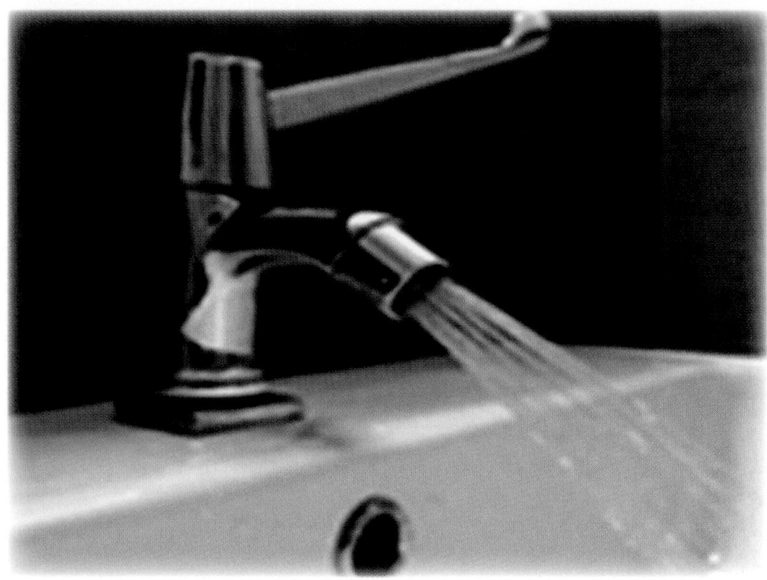

Figure **18** A water-saving aerator

they are used. Existing taps can be retrofitted to deliver water more efficiently. These aerators, which are not very expensive, are designed to dispense 2–8 litres of water a minute.

Buy low-flow taps when you are changing taps or faucets in your home. These are not expensive and can save up to 80% water compared to the conventional taps.

Whereas regular showers deliver 15–20 litres per minute, water-saving showers (Figure 19) typically deliver only 6–8 litres or even less without compromising on the bathing experience. These devices restrict and limit the amount of water that is let out of an existing tap or shower and are recommended in situations where it is not feasible to install aerators. You can save up to 60% of water flow by using a flow restrictor.

A dual-flushing water-efficient toilet uses only 25% of the water used by conventional flushing systems (Figure 20).

Figure **19** A water-saving shower and parts for retrofitting

Figure **20** A water-efficient dual-flushing toilet

Water quality is as important as quantity

The quality of water we drink or use for our daily needs varies
from place to place and depends largely on the source of water, and
natural sources of water have been increasingly threatened with

- High nitrate content in water can impair oxygen transport in infants, and nitrates, if consumed, can impair neuromotor functioning and development.
- High levels of arsenic, lead, asbestos, cyanides, copper, etc. in water can be extremely harmful to health and can lead to many problems including dental and skeletal fluorosis due to excessive levels of fluoride, arsenicosis due to high levels of arsenic, and endocrine disruptions and neurological damage due to excessive levels of mercury.
- Biological contamination of water due to viruses, bacteria and protozoa can lead to a range of waterborne diseases such as diarrhoea, dysentery, cholera, typhoid, and hepatitis that can prove fatal if not treated in time.
- Contaminated water and poor sanitation are linked to transmission of cholera, diarrhoea, hepatitis A, typhoid, and polio.
- Microbial contamination, for example the presence of *Escherichia coli*, is a serious problem and an important indicator of water quality.
- Chemicals from industrial effluent can leach into groundwater and have a long-term impact on people living in the vicinity.
- Chemicals from farmlands include pesticides, nitrates, and phosphorus, which can contaminate both groundwater and surface water, leading to adverse effects on health.
- Human settlements also release harmful chemicals into water. These chemicals include chlorinated solvents, petroleum hydrocarbons, benzene, nitrate, ammonia, and heavy metals—all with adverse long-term impact on health.
- Chemical associated with disinfecting water may combine with natural organic matter and generate by-products such as trihalomethanes and haloacetic acids. These can cause cancer and diseases of the reproductive system.

contamination from industrial, agricultural, and biological
activities. Contaminated water poses a serious health risk to
infants and adults.

For our daily needs of water, we rely mostly on municipal
supply and sometimes on unregulated – and often illegal – sources
such as groundwater. And because we are no longer confident of
the quality of tap water, we use water purifiers for pre-treatment
of water for drinking, bathing, and cooking.

Groundwater in several regions of India is highly polluted
with fluorides, arsenic, nitrates, iron, and heavy metals and also
from residues of pesticides and fertilizers used on farms. Toxic
substances from untreated industrial wastes and landfills as well
as bacterial contamination from surface soil and water can also
contaminate groundwater.

Common contaminants occurring in drinking water can be
classified into several categories.

- Inorganic contaminants
- Organic contaminants
- Biological contaminants
- Radiological contaminants

Choice of appropriate treatment techniques is necessary to
avoid hazards related to water quality. The treatment techniques
are not included in the scope of the booklet.

Energy

About one-third of global energy is consumed in homes and
in commercial and public buildings, collectively referred to as
buildings. Energy is used for cooling or heating spaces, lighting,
ventilation, refrigeration, for a variety of electrical appliances, for
heating water, cooking, and so on. **According to the International
Energy Agency, among all regions of the world, energy
consumption in buildings is expected to grow the fastest up
to the year 2040 in India.** Also, within the buildings sector, it is
homes, or the residential sector, that consume the maximum

o Air conditioning consumes the highest energy in the buildings sector.

o After air conditioning, homes account for the highest energy consumption, with heating water and lighting at the top, followed by electrical appliances.

o In office and other commercial buildings, after air conditioning, it is lighting that accounts for maximum energy consumption.

o Energy consumption for heating or cooling enclosed spaces is greatly influenced by the design and materials used for walls, windows, and roof; therefore, these components need to be chosen to suit the local climate.

o All electrical appliances, especially TV sets and air conditioners, need to be switched off when not in use—not through remote controls, because they keep using up electricity while in stand-by mode, but from the mains. Even low-power gadgets such as chargers, adaptors, and invertors consume substantial amounts of electricity while in stand-by mode.

o The above also applies to desktop computers and laptops: in many offices, these devices are never switched off completely – typically, only the monitor is switched off – which is why electricity consumption in office buildings is high even outside working hours.

energy—and therefore also offer huge opportunities to save energy.

A lot can be done even with existing buildings although, ideally, energy efficiency should be taken into account from the earliest stages in the architect's office, when a building is in its conceptual stage. **This section shows how you can save energy** at the design stage and also if you occupy an apartment or an independent home already constructed.

Climate-Responsive Design

As emphasized repeatedly, buildings must be designed to suite the local climate so as to achieve the required levels of thermal comfort and lighting from the design itself, thereby minimizing the dependence on appliances. For example, Chennai has a warm and humid climate and therefore benefits from lightweight structures that do not store heat. Jaipur, on the other hand, has a hot and dry climate, making heavy structures, such as thick walls, more appropriate because they absorb and retain heat during the day and dissipate it outwards, when the outside temperatures are lower than those within in the evenings. Day and night temperatures fluctuate widely in hot and dry climates, and buildings need to be constructed to cope with such fluctuations.

Therefore, before building a home, consider the local climate; if buying a home already constructed, opt for a unit the design of which is climate responsive and has integrated in the design as many of the features discussed here as possible.

Energy efficiency

Climate-responsive building design is the first step; the second step is to integrate the most energy-efficient equipment for air conditioning, lighting, and other end uses so that comfort is achieved at minimal or optimal cost in terms of energy.

Integrating renewable energy with building design
Three major forms of renewable energy can be integrated into buildings design to meet the building's energy demand, either partly or fully. The three forms are solar energy, wind energy, and energy from waste, especially biogas. You can even sell the energy generated through these means to the utility company that supplies electricity to your area (https://pmjandhanyojana.co.in/sristi-scheme-rooftop-solar-power-plant-subsidy-government/).

If you are planning to install solar panels on the roof of an existing building, see *Sun Through the Roof*, which, along with the present book, is part of the 'Concerned Citizen' series of short books.

Using building materials and construction with low embodied energy

In attempting to save energy through appropriate construction materials, we must not lose sight of the fact that it also takes energy to produce those materials in the first place and to transport them to the site at which they are used. This aspect is referred to as 'embodied energy'. Materials and construction technologies that consume less resources are considered 'greener' and more sustainable because their embodied energy is low, as are locally available and locally manufactured products, because less energy is spent in transporting them to the site.

Besides design and construction materials, internal finishing and furnishing including paints and fixtures (plumbing and electrical fittings, for example) also need to be considered.

Conventional paints, sealants, and adhesives used for interiors of a building are formulated with solvents that emit VOC. These are toxic chemicals that, when they evaporate, pollute indoor air and affect the health of the occupants both in short term and long term. Therefore, use paints, adhesives, and sealants that have no VOC or only low levels of VOC.

Also, for the interiors, use materials that have low embodied energy, preferably materials that are based on renewable sources, environment-friendly, and non-toxic. Bamboo is a rapidly renewable resource and can be used effectively to construct structures or as a reinforcement. Use of industrial wastes such as fly ash in making building blocks not only uses up waste from thermal power plants but also helps in saving precious topsoil. Aerated concrete blocks use fly ash as a key ingredient and provide insulation that can control heat gain.

Designing for different climate zones of India

India has five climate zones, namely hot and dry, warm and humid, composite, temperate, and cold (Figure 21).

Let us see some simple solar passive or climate-responsive design strategies to reduce the energy used for space cooling or heating and for lighting.

Orientation

Optimum orientation (Figure 22) refers to the orientation of the longer axis of a building such that the building receives maximum solar radiation in winters and minimum solar radiation in

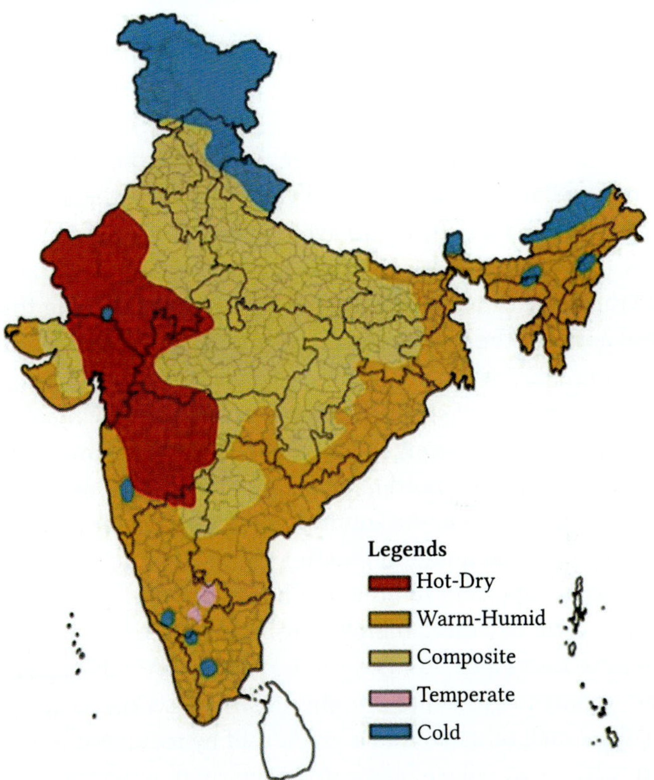

Figure **21** The five climatic zones of India

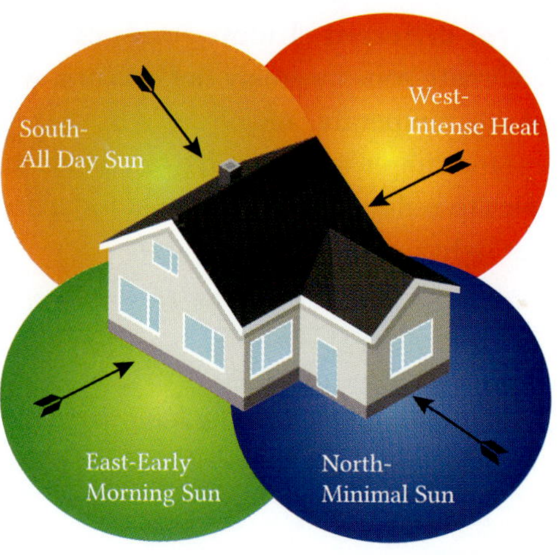

Figure **22** Optimal orientation of buildings
with respect to solar radiation

summers. Parts of a building that face the east or the west receive
high solar radiation throughout the year; therefore, these facades
need to be designed critically to avoid overheating during summer
months. The total surface area these facades present to the sun
should therefore be as small as possible. Parts of a building that
face the north or the south receive solar radiation at low intensity.
The southern facade, however, needs special treatment during
summer in the form of shading devices. However, in winters,
south-facing surfaces receive radiation at the highest intensity;
therefore these facades need to be larger. (This is true of the
northern hemisphere; in the southern hemisphere, north-facing
surfaces receive maximum radiation.)

It is also important to note that horizontal surfaces receive
maximum solar radiation; therefore roofs need special attention
to minimize heat gains. Roofs could be shaded (a double roof, as
it were), painted with white or light finishes (also known as cool
roofs), insulated, or made into a green roof by making it into a
garden, which can reduce heat gains from roofs substantially,
thereby making the top floor of the building more comfortable.

For individual homes

o Orient the longer facades along the north–south axis. Maximize openings along that axis.
o Design shading devices consistent with the path of the sun.
o Orient smaller facades along the east–west axis. Minimize openings along that axis to lower space-cooling requirements.
o Treat or design roofs to reduce external heat gains from the sun's radiation.

For apartment homes

o Select homes with long facades oriented along the north–south axis.
o Make sure the windows are well shaded or recessed to cut down high solar radiation from the east or the west.
o If windows are not shaded, install movable external shading devices on critical facades, mostly south- and west-facing windows. Movable shading devices can be adjusted as required and thus allow occupants to enjoy diffuse light or views from the window when the sun is less intense.
o If your home is on the top floor, insist that the roof be either insulated or shaded or its heat gain minimized by other means.

Landscaping

Landscaping can modify the microclimate of a site, and appropriate landscaping can help in achieving thermal comfort outdoors and, in turn, indoors as well. For example, tall and leafy trees, by reducing direct solar radiation striking the built-up surfaces, can keep buildings and paved areas cooler. Shade from trees and evapotranspiration from vegetation lower the temperature of ambient air of the microclimate, as demonstrated in the study mentioned earlier (Figure 9), which shows the difference in temperatures between the green campus of the Indian Institute of Science and the Electronic City, Bengaluru, India, with its sparse vegetation and buildings dominated by steel and glass.

Tips on landscaping

o In hot and dry climates, landscaping, through vegetation and water bodies, helps to lower the temperature of ambient air. Species native to the region are preferred because they require less maintenance.

o In composite climates, landscaping using deciduous trees provides shade in summers and, because the trees shed their leaves in winter, allows ample sunlight in winters.

o In warm and humid climates, trees provide shade and may also increase wind speeds by their funnelling effect, which contributes to thermal comfort outdoors and indoors.

Form of the building

The microclimate within a building is affected by many aspects of its design including its form, shape, the number of facades and their orientation, and the ratio of its surface area to its volume (S/V ratio). Compact forms, for example, are preferred in cold or hot climate because their low S/V ratio lowers heat gain and heat loss, whereas a higher S/V ratio is desirable in warm and humid climates to create more airy spaces.

The scarcity of land has compelled cities to expand skywards, making high-rise buildings increasingly common. High-rise or tall building forms interact with the microclimate very differently from what low-rise buildings do. In high-rise buildings, the availability of sunlight increases with height (Figure 23) as does the speed of wind. High-rise buildings also cast long shadows, which is why it is important to check the presence of solar installations in the neighbourhood.

In green-buildings projects, the distance between two adjacent high-rise buildings is maintained such that even the lowest floors receive the recommended levels of natural daylight. The Bureau of Indian Standards recommends that these levels be the highest in the kitchen. Kitchens should also have external windows, which not only let in daylight but also improve the quality of air in the kitchen: dark or windowless kitchens not

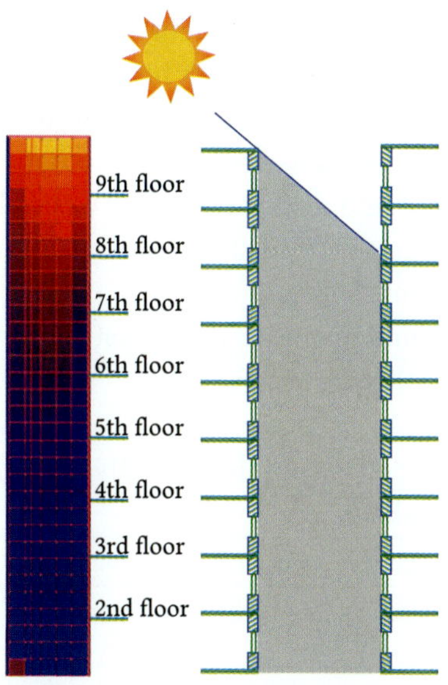

Figure **23** High-rise building with low solar radiation on lower floors and high radiation on higher floors. The building envelope is treated differently on higher floors in keeping with high solar radiation.

only consume more energy on lighting but are also less hygienic because moulds thrive in dark, stuffy, and humid places.

One may also change the size of the windows depending on the floor: windows on higher floors could be smaller because those floors get higher levels of daylight and higher wind speeds.

Similar to the principles of 'vaastu', following the principles of green buildings can bring good health and prosperity.

Building envelope

A building envelope, or the skin of the building, comprises external walls, windows, and the roof. These components determine the amount of light, wind, and heat transfer from the outside to the inside of a building.

External walls are a major component of the building envelope, and their capacity to store and conduct heat, thickness, finish, and the material used in their construction have a huge impact on the number of comfortable hours as well as on the requirements for air conditioning. In hot and dry climates, for example, construction materials with higher capacity to store heat are important in maintaining indoor conditions comfortable.

Table 2 suggests some materials and construction technologies appropriate to different climates.

Table **2** Recommended materials for walls for different climatic zones in India

Hot and dry	Warm and humid	Composite	Moderate	Cold
Brick walls	Hollow concrete blocks	Brick walls	Shaded walls	Brick walls
Stabilized mud blocks	Autoclaved aerated concrete blocks	Stabilized mud blocks	Insulated walls	Stabilized mud blocks
Solid concrete blocks	Insulated walls	Solid concrete blocks	Brick walls	Solid concrete blocks
Flash ash bricks	Lightweight construction	Fly ash bricks	Stabilized mud blocks	Insulated walls
Cavity walls		Cavity walls, insulated walls	Solid concrete blocks, fly ash bricks	

Table 3 gives the heat transfer coefficient for different materials used for walls. The coefficient is a measure of how well a structure conducts heat and is a function of its area, thickness, thermal conductivity, and so on.

Tips for building owners
In naturally ventilated buildings, walls with high thermal capacity (higher ability to store heat) delay the transfer of heat from outside to inside in hot climates and vice versa in cold

Table **3** Heat transfer coefficient for different materials

Material	Heat transfer coefficient (W/m²K)
Clay bricks	2.13
Fly ash bricks	1.9
Solid concrete blocks	2.7
Hollow concrete blocks	1.89
Single-glazed units	5.0–6.0

climates. Such walls, including those of solid concrete, brick walls, and mud walls are advantageous in both hot and cold climates. On the other hand, lightweight construction and materials that do not store heat are recommended for warm and humid climates.

Insulation of walls is recommended if buildings are air conditioned. Insulation is like a woollen blanket or a sweater: it does not allow body heat to escape and keeps us warm—and it also keeps out the heat from the outside.

In existing homes, if east- and west-facing walls are not treated to reduce external heat gains, they need to be converted into green walls with natural vegetation to shade them and also to cool them through evapotranspiration.

Window design

Windows and glazed areas allow light, heat, and wind as well as afford a view of the outside. Among all the other components of a building envelope, windows contribute the most to heat gain.

Consider the thermal properties of walls and those of windows with single-glazed units. Glass lets in more heat, 2 to 3 times that let in by walls (Figure 24). Therefore, the proportion of glazed area should be greater in cold climates and smaller in hot climates.

For new office buildings

◐ Window size and positioning should be designed depending on the climate and the location of the building.

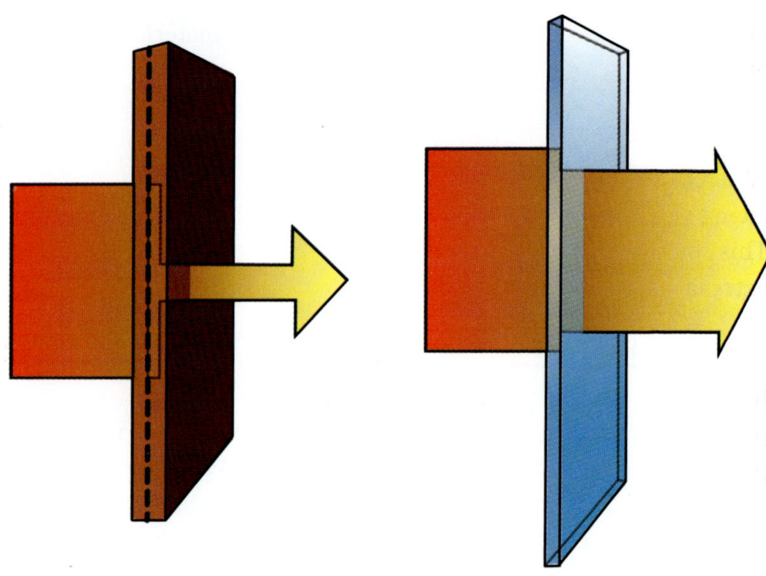

Figure **24** Higher heat transfer through windows than through walls

o Windows that are large enough to account for about 20%–30% of the total area of the wall are considered adequate for a predominantly hot climate in that they let in enough of natural daylight without letting in too much of heat.
o In cold climates, that proportion can go as high as 60% of the total wall area.

For retrofit of existing office buildings into efficient office building
o If windows are too large, leading to excess daylight, which causes glare, and overheating, retrofit the windows using either advanced glazing systems or external shading devices. Several types of coatings are available in the market, which control the amount of solar heat and light inside the buildings. For air-conditioned buildings, replace single-glazed units with double-glazed units. But while choosing a film, ensure that it allows daylight to come

in while blocking heat. Ask your provider for a film that has
higher light transmission but a low solar factor.

For new homes
⊙ Whereas large openings are recommended for north- and south-
facing walls, windows facing east and west should be well shaded.
This could be achieved through pergolas, balconies, canopies, or
external shading devices.

For retrofit of existing homes into efficient homes
⊙ If rooms are overheated because of faulty window design, it
is important to consider external shading devices to reduce
dependence on air conditioning for thermal comfort.

Shading devices

Shading devices are advisable for all climate zones. In hot
climates, external shading devices help in cutting down direct
solar heat gain as well as help in controlling glare. They also
help in keeping the rain out. It is important to mention here that
internal louvers are not as effective as external shading devices in
reducing heat gain and glare.

Ideally, shading devices should be designed by architects
based on the sun path diagram of the building location. It is a
good practice to design fixed external shading devices for north-
and south-facing walls and, if economically viable, movable
shading devices for east- and west-facing walls. This allows one to
enjoy the view to the east once the sun moves towards the west,
typically after 1 p.m.

Roof design

The roof, being horizontal, receives the highest solar radiation
intensity (Figure 26). This is the reason, living spaces adjacent to
external roof are warmer than lower floors. Air conditioning load as
well as discomfort hours could be higher on upper floors if the roof
is not designed taking into consideration the climate of the city.

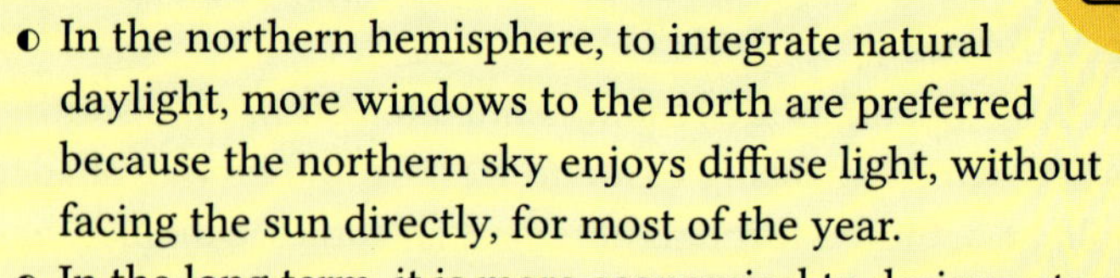

- In the northern hemisphere, to integrate natural daylight, more windows to the north are preferred because the northern sky enjoys diffuse light, without facing the sun directly, for most of the year.
- In the long term, it is more economical to design external shading devices with single-glazed windows than using glazing that controls how much of sunlight is let in (Figure 25).

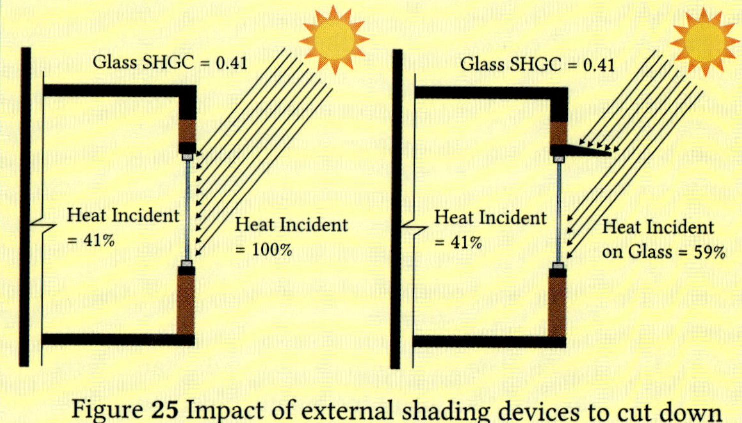

Figure **25** Impact of external shading devices to cut down external heat gains

Different types of roofs can be used depending on the climate. **A few types of roofs are described here**.

Cool roofs Just as light-coloured clothes keep us cool during summer, cool roofs, by reflecting more sunlight, keep the insides of a building cooler than conventional roofs can. As can be seen from Figure 27, surface temperatures of a cool white roof (Figure 28) can be as much as 30 °C lower than those of a conventional cement-tiled roof. The lower the temperature of an

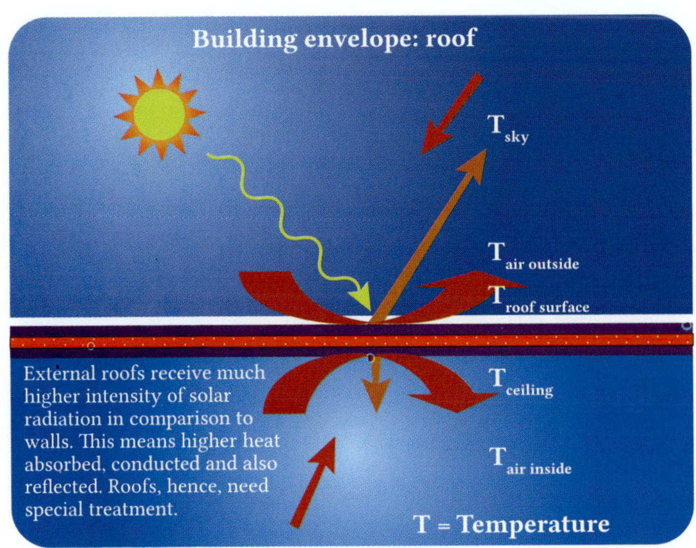

Figure **26** Heating of roof by direct incident radiation

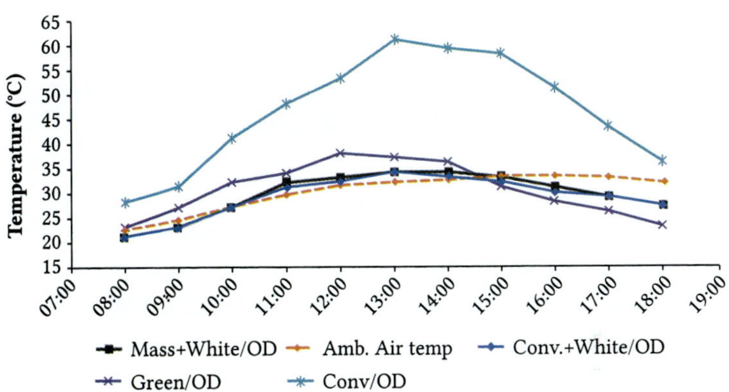

Figure **27** Hourly surface temperatures of roofs of different finishes in summer (2 April) in Bengaluru

external surface, the lower the heat gains from roof, making top floors adjacent to roof thermally comfortable.

Figure **28** A cool white roof being installed

Green roofs Using the roof as a base for a rooftop garden has a moderating influence on the microclimate within: the soil insulates the roof, and the plants, through evapotranspiration, help in lowering the temperature of the air layer immediately above the roof. Living spaces below such green roofs (Figure 29) are particularly comfortable.

Shaded roofs Having a shade over the roof keeps it cooler by avoiding exposure to direct solar radiation.

Insulated roofs External insulation, for example in the form of a layer of insulating material to cover the roof slab, avoids heat being stored in the building mass, thereby reducing the transfer of heat from the outside to the inside and, in turn, reducing the load on air conditioning and keeping the insides comfortable for longer hours.

Tips for building owners
- For new buildings, choose any of the above types of roof.
- For existing buildings without any of the roofs described above, retrofit the roof with a cool roof by painting it white with a reflective coating. This is the most doable option without making any structural changes to the existing roof.

Figure **29** A green roof to keep the inside air warmer in cold climate and cooler in hot climate

Cooling techniques that consume less energy
Despite both temperature and relative humidity being high inside, air movement around occupants can help in making them feel comfortable.

In energy-efficient green buildings, it is important to analyse the climate of the city and note the time of the year during which natural ventilation alone can be adequate. This is usually possible when the conditions outside are more favourable than the conditions inside.

Air movement inside buildings can be increased by creating a pressure difference. Air moves from regions of higher or positive pressure to those of lower or negative pressure. A pressure gradient can be achieved by creating openings on both the windward side and the leeward side. If windows account for 15%–20% of total wall area, air speed within is approximately 1.2 times the speed outside (Figure 30).

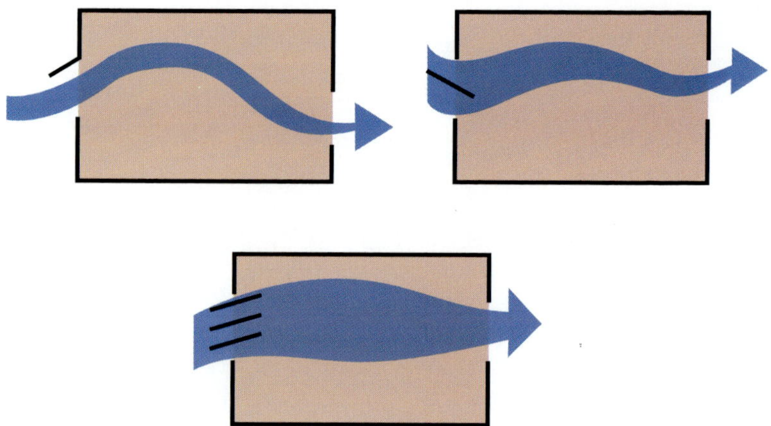

Figure **30** Natural ventilation achieved through difference in air pressure

Natural ventilation
Ventilation can also be enhanced using tall but narrow spaces by exploiting the stack effect driven by buoyancy. These tall spaces include atriums, shafts, and stairwells inside a building. Such spaces trap a column of air with warmer air at the top and cooler air at the bottom. The warm air within escapes through the openings at higher levels of the stack (column), while cooler air rushes in through the openings at lower levels of the stack. This ensures a current of air. Ventilation using the stack effect is used effectively in many office and institutional buildings.

Solar chimneys To exploit the stack effect and the pressure difference more effectively, solar chimneys could be introduced at higher levels of tall spaces (Figure 31). Such chimneys make the hot air escape even faster, thereby ensuring an overall faster turnover. Solar chimneys, using natural ventilation, facilitate the movement of air within buildings in hot climates. Ideally, such chimneys should be placed to the south or southwest because these aspects are warmer. The chimneys could also be separate structures but connected to the main building.

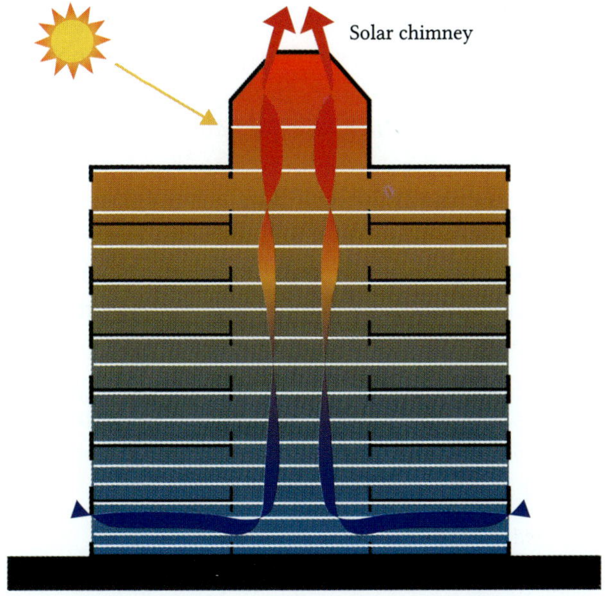

Figure **31** Natural ventilation by exploiting the stack effect

In the future, 'hybrid' buildings are likely to be more common, which use both natural ventilation and ventilation secured through such means as air conditioning and exhaust fans, for example. In designing ventilation, it is important to consider such challenges as noise, the quality of outside air, and dust.

Evaporative cooling
When the air is hot and dry, evaporative cooling lowers air temperature through evaporation of water and thereby cools the living spaces within a building. Water bodies such as ponds, lakes, and fountains in a courtyard help achieve thermal comfort in hot and dry climates. Mist cooling and 'passive down draft evaporative cooling' are two more ways of exploiting evaporative cooling within a building. Mist cooling involves forcing water under pressure through fine nozzles: the droplets are too small (smaller than 50 μm) to settle on surfaces to

wet them; instead, they evaporate quickly and bring down air temperature. Passive down draft evaporative cooling is the opposite of the air chimney in that a layer of cold air is created at the top of a stack; the cold air, being heavier, travels downwards, and is then introduced into living spaces at lower levels. Another commonly used devise is the desert cooler (Figure 32), which comprises, enclosed within a box, stored water, evaporative pads, a fan, and a pump: the stored water is pumped and released above the evaporative pads; the water percolates down and wets the pad; while the fan pushes the air within the box to the outside, outside air enters the box and is cooled as it passes through the wet pads. The system works very well in hot and dry climates and also during the hot and dry months in the composite climate.

Tips for building owners
- Evaporative cooling requires water and therefore may not be feasible if water is scarce.
- Evaporative cooling is not effective if the outside air is humid, because high humidity slows down evaporation.

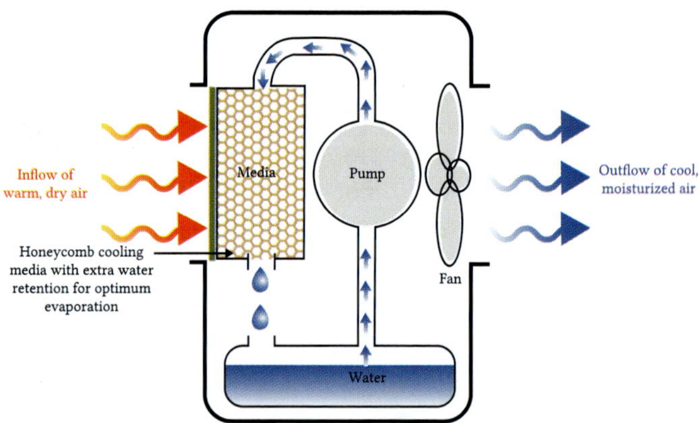

Figure **32** Working of an evaporative cooler

Geothermal cooling

In geothermal cooling, a layer of earth (soil) acts as an insulating layer. At a depth of 4 metres, the temperature remains fairly constant—which is why basements are cooler in summer and warmer in winter. Air temperature at that depth is close to the average temperature of the place. Air inside tunnels or pipes embedded at that depth, or deeper, is of the same temperature as the surrounding earth, and ambient air passing through these pipes and then introduced into living spaces makes them cool in summer and warm in winter.

Heat pumps

A heat pump is a system used to heat or cool an enclosed space or small quantities of water by making thermal energy flow from a cooler space to a warmer space using the refrigeration cycle, moving heat in the direction opposite to that in which heat would have travelled in the absence of external power. Heat pumps transfer heat energy to a thermal reservoir. Different kinds of heat pumps are used in buildings for cooling or heating enclosed spaces.

Air-source heat pumps extract heat from ambient air and are widely used for heating water in homes. In the process of heating the water, the surrounding air gets cooled, which is then used for space cooling. Such systems are 3–4 times more efficient than conventional geysers.

Water-source heat pumps are among the most efficient HVAC (heating, ventilation, and air conditioning) equipment for cooling and heating. Although the working principle of these pumps is similar to that of air-source heat pumps, water-source pumps use a water loop instead of air to extract heat and to supply hot water or to heat living spaces: while one side of the heat pump generates hot water, the other side generates cold water, which can be used for space cooling. These pumps are therefore particularly suitable for simultaneous heating and cooling or when warm water is available (waste heat) and are also viable and efficient in cold regions: the water loop extracts heat from water bodies in winter because the water is warmer than the surrounding air or soil.

Ground-source heat pumps use the earth as a heat source in winters and as heat sink in summers. In winter, the earth's heat is transferred to a building through some liquid that circulates through an underground network of pipes. In summer, heat from a building is transferred to the earth, which acts as a sink, using the same network of pipes. For more details, visit <https://vikaspedia. in/energy/energy-production/ground-source-heat-pumps>.

Tips for building owners
- Geothermal cooling can be for homes and offices as well as for institutional buildings but likely to be feasible only for smaller projects.
- Earth air tunnels can provide space cooling and can also be used to pre-cool the air before using it for refrigeration, thereby lowering the load of refrigeration.
- It is advisable to integrate geothermal cooling with solar chimneys. Solar chimneys help warm air from enclosed spaces to rise and escape to the outside, thereby creating a current of air for the cooler air from the underground tunnels to enter the same spaces.

Radiant cooling
Cold slabs or ceilings that enclose pipes through which chilled water is circulated absorb the heat (radiation) generated by people, lights, equipment, etc. close to such slabs or ceilings. Spaces are cooled because the ceiling slab is cooler than the ambient air, thus taking heat away from the living space, cooling it in the process. This method is a more efficient way of cooling than air conditioning, which is the conventional method of space cooling. However, radiant cooling takes care only of 'sensible' heat but can neither lower humidity nor take away 'latent' heat, and does not circulate air either. ('Sensible' heat is heat that can be

sensed and measured using a thermometer; 'latent' heat is hidden or invisible and cannot be measured using a thermometer—which will continue to show the temperature at which water is boiling, for example, but cannot account for the heat that is converting that water into vapour.) However, fresh air is essential for good health and well-being of occupants, which is why fresh air is introduced and humidity is regulated through a dedicated outdoor air system.

Radiant cooling or heating can work in all climate zones and is 30% or more efficient than air conditioning. Radiant cooling also provides higher-quality indoor air and comfort cooling to occupants.

Table 4 summarizes the aspects of climate-responsive design discussed so far appropriate to the five climatic zones in India.

Energy-efficiency measures

Energy performance or intensity is usually defined as units (kilowatt-hours) of energy consumed annually per unit area (square metres)—the lower this number, the higher the efficiency of the building. Energy consumption of buildings varies a great deal depending on the type of the building, number of hours in a day the building is used or occupied, efficiency of equipment, architectural design, climate, and so on. For example, an air-conditioned office building in Chennai (warm and humid climate) will probably have a higher energy performance than a similar building in Bengaluru (moderate climate). Highly energy-efficient residential and commercial buildings have a total energy intensity of 50–100 kWh per square metre of floor area per year, compared to a typical value of 200–400 kWh (GEA, 2012: *Global Energy Assessment: Towards a Sustainable Future*, Cambridge University Press and IIASA, Austria). The first step to achieve energy efficiency in buildings is to design a climate-responsive building, which is discussed in the above section. The second step is to design and select energy-efficient systems, mainly lighting, space conditioning, and appliances as discussed below.

Table **4** Summary of climate-responsive measures for different climate zones in India

Measure	Composite	Hot-dry	Warm-humid	Moderate	Cold
Orientation					
Long facades, oriented north–south	☑	☑	☐	☑	
Landscaping	☑	☑	☑	☑	
Building form (low surface-to-volume ratio)	☑	☑		☑	☑
Building envelope					
Walls					
High thermal mass	☑	☑		☑	☑
Insulated walls	☑	☑	☑	☑	☑
Lightweight construction			☑		
Windows					
Windows-to-wall ratio lower than 40%	☑	☑	☑	☑	
More or larger windows on north and south; fewer or smaller openings on east and west	☑	☑	☑	☑	
Windows-to-wall ratio more than 40%					☑
Double-glazed windows for air-conditioned spaces	☑	☑	☑	☑	☑
Single-glazed windows with external shading for natural ventilated buildings	☑	☑	☑	☑	
External shading devices	☑	☑	☑	☑	☑
Daylight integration	☑	☑	☑	☑	☑
Green roof	☑	☑	☑	☑	☑
Cool roof	☑	☑	☑	☑	
Insulated roof	☑	☑	☑	☑	☑
Shaded roof	☑	☑	☑	☑	
Natural ventilation	☑	☑	☑	☑	☑
Evaporative cooling	☑	☑			
Geothermal cooling or heating	☑	☑		☑	☑
Radiant cooling or heating	☑	☑	☑	☑	☑

Energy-efficient lighting
Lighting accounts for about 20% of the total electricity
consumption in a building, and energy-efficient lighting seeks to
provide visual comfort with minimum use of energy, thereby not
only saving money but also contributing to reducing the emissions
of greenhouse gases from the buildings sector. Energy-efficient
lighting is achieved through the following measures.
o Integrating daylight with artificial lighting
o Integrating daylight with 'smart' (controlled) lighting
o Matching the level of illumination (expressed in lux) to the task
o Using efficient lamps and fixtures
o Using lighter colours for interiors

In India, the Energy Conservation Building Code (ECBC)
recommends that natural daylight be integrated with artificial
lighting through appropriate controls. **When glare-free daylight
is available, artificial light is no longer needed, which is why
luminaires in areas served by daylight should be equipped
with manual or automatic controls to switch them off the
luminaires altogether or to lower the brightness as required**.
(A lamp is different from a luminaire in being only a part of the
latter: a luminaire refers to the complete assembly comprising
the lamp and other parts connected with the lamp to distribute,
position, and protect it.)

Recommended levels of illumination
In green and energy-efficient buildings, efficiency is achieved
by matching the level of illumination – as recommended in the
National Building Code – to the task performed in a particular
space or setting. The recommended illumination levels in Figure 33
should be met at all times whenever the space is being used for the
stipulated task. The levels are stipulated for the task plane, which
is taken as 850 mm above the floor level. The distance is specified
because, as discussed earlier, the level of illumination, measured in
lux, is also affected by the distance from the source of light.

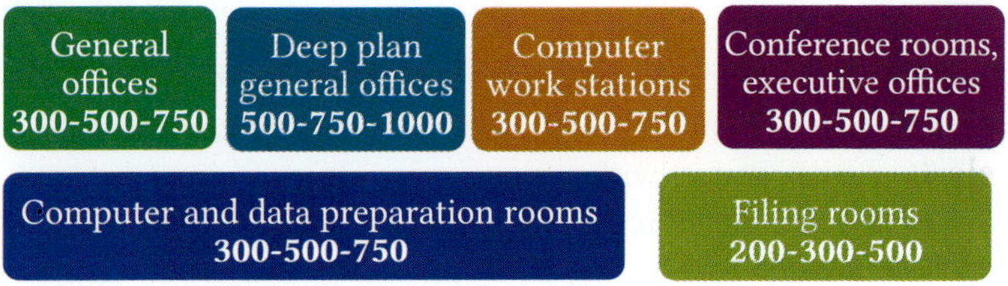

Figure **33** Recommended levels (in lux) of illumination for different tasks in a commercial setting

Efficient lighting fixtures

In ensuring the required level of illumination, the energy expended for the purpose should also be considered. Accordingly, the Energy Conservation Building Code stipulates the rating (in watts) of the luminaires per unit area of the surface being illuminated (Table 5). The installed wattage of the luminaires – including lamps, ballasts, regulators and central devices – must be within the maximum permissible power densities specified for interior lighting of various building types. **The efficiency of lighting design depends upon the selection of the**

Table **5**   Maximum permissible lighting power densities recommended by building code

Space function	LPD (W/m²)	Space function	LPD (W/m²)
Office, enclosed	11.8	Library	
Office, open plan	11.8	Card file & cataloging	11.8
Conference/meeting/ multipurpose	14.0	Stacks	18.3
Classroom/lecture/ training	15.1	Reading area	12.9
Lobby	14.0	Hospital	
For hotel	11.8	Emergency	29.1

luminaire and the lamp. The luminaire should be selected using available photometry data, should ensure uniform distribution of light, control glare, and ensure the necessary electrical safety.

Lamps should be selected keeping in mind their luminous efficacy and CRI (colour rendering index). Luminous efficacy of a light source is a measure of how efficiently it produces visible light and is expressed at the ratio of the output of light (in lumens) to the power rating (in watts) of the source of light (Table 6).

Table **6** Luminous efficacy for different types of lamps

Product type	Typical luminous efficacy
LED cool white package	130
LED warm white package	93
LED A19 lamp (warm white)	64
LED PAR38 lamp (warm white)	52.5
High intensity discharge (high watt)	
Lamp and ballast	120:111
Linear fluorescent	
Lamp and ballast	118:108
High intensity discharge (low watt)	
Lamp and ballast	104:97
Compact fluorescent lamp	63
Halogen	20
Incandescent	15

Efficient space conditioning

The main purpose of buildings is to provide a comfortable and secure indoor environment to its occupants. Thermal comfort is a particularly important part of such environment because it affects the well-being and productivity of occupants. Green buildings deploy climate-responsive architecture in their design and may also integrate passive cooling or heating technologies in the design. However, even these buildings may have to depend on air conditioning to provide thermal comfort if the outside environment is unfavourable, if pollution levels outside are very high, or if the demand for space cooling or heating is high. Let us therefore discuss ways of achieving such comfort while minimizing energy consumption.

Question and answers on air conditioning
Q. *Why is it important to optimize energy demand for cooling and heating?*
Air conditioning systems use refrigerants, which have adverse impacts on global warming and lead to higher emissions of greenhouse gases. Optimizing air conditioning loads would reduce the extent of refrigeration to be used in a building, thereby lowering its negative impact on the environment.

Q. *What are the adaptive set points or range for air-conditioned buildings for thermal comfort in India? How is this information helpful?*
Most often, the set temperature is too low, which makes excessive – and avoidable – demands on air conditioning; raising the set temperature even by a few degrees can make a great difference. India's National Building Code, developed by the Bureau of Indian Standards, maintains that thermal comfort lies between 25 °C and 30 °C, the midpoint of that range, namely 27.5 °C, being optimal.

However, for air-conditioned buildings, the specifications are more stringent. For example, for most spaces, the recommended setting is a DBT (dry-bulb temperature) between 23 °C and 26 °C and relative humidity between 55% and 60%.

It is now up to you to think: Would you like to increase the set temperature to 27.5 °C or even to 30 °C, depending upon what you do within that space and without compromising on thermal

comfort? With this simple change, you would be doing your bit
for the environment.

Q. *How do we measure the efficiency of an air-conditioning system?*
The efficiency of an air-conditioning unit can be assessed
either in terms of its EER (energy efficiency ratio) or as its CoP
(coefficient of performance). The efficiency ratio is the ratio
of the net cooling capacity (in kilowatts) to its total rating (in
watts) under the specified operating conditions for which the
unit is designed. The building code recommends the EER for
unitary, split, and packaged air-conditioning units. The Bureau
of Energy Efficiency rates the efficiencies of various appliances.
The efficiencies are indicated in terms of the number of stars.
The air-conditioning units that have earned such stars from the
BEE are certified, and their EERs are part of the labelling. The
higher the ratio, the higher the efficiency.

The coefficient of performance is the ratio of amount of heat
removed to the rate of energy input, in consistent units, for a
complete refrigeration system or a specific portion of the system.
The building code also stipulates minimum acceptable values of
the CoP for air-cooled and water-cooled chiller systems.

Types of air-conditioning systems
The most common air-conditioning systems are convective
cooling or heating systems. These can be further divided into
unitary systems and central systems. The unitary systems are
typically stand-alone units of smaller capacities whereas large-
capacity systems are usually centralized systems.

• *Unitary systems* Window air conditioners and split units
are the most commonly used air-conditioning units: the former
are typically available with cooling capacities of 0.5 to 2.0 TR
(tonnes of refrigeration) whereas split units are available in
capacities ranging from 0.5 TR to 5 TR. In choosing a unit of

the right capacity, we need to take into account the size of the room to be cooled and the external environment. A unit with excess capacity means wastefully high electricity consumption and thus a higher adverse impact on the environment. Split air conditioners based on inverter technology, because they use a variable-speed compressor, are more efficient than conventional split conditioners: they are also better at maintaining the set room temperature and provide faster cooling than conventional air conditioning units do. For more details on this technology, see <https://www.beeindia.in/inverter-ac-technology/>.

A 1.5-tonne split unit that has earned a 5-star rating from the BEE will consume 30% less energy than that consumed by a unit that has earned only 1 star (Table 7).

Table 7 Split air-conditioning units with 1 star and 5 stars compared

NOTE Units are run 8 hours a day for 5 months in a year and 1 unit (1 kW) of electricity costs Rs 4.

Parameter	Conventional technology		Inverter technology	
Star rating	1 star	5 stars	1 star	5 stars
Energy efficiency ratio	2.5	3.5	2.5	4.5
Cooling capacity (watts)	5200	5200	5200	5200
Input power (watts)	2080	1486	2080	1155
Units consumed per day	17	12	17	9
Monthly cost of electricity (rupees)	2040	1440	2040	1080
Annual savings (rupees)	0	3000	0	11,520

SOURCE https://www.beestarlabel.com/Content/Files/Session_2.pdf

Thus a 5-star 1.5-tonne split unit with invertor technology will consume 45% less energy than that consumed by a conventional unit that has earned only 1 star. A 3-bedroom home with three such units (with inverter technology and a 5-star rating) can save Rs 34,560 a year.

• *Central air-conditioning systems* Large buildings have large cooling demands, which are met through large air-conditioning plants, which are mainly of two types, namely air cooled and water cooled. In both types, the refrigeration cycle occurs inside of a chiller. In air-cooled chillers, heat is led out directly to the outside, whereas in water-cooled chillers, water pipes run on the condenser side of the chiller and are led inside a cooling tower. Heat from the chiller is also led inside the cooling tower. Water is sprayed inside the tower to cool the air inside the tower. This system makes water-cooled chillers more efficient than air-cooled chillers.

The building code also stipulates efficiencies for centralized air-conditioning units; these stipulations should be followed to achieve energy efficiency and to save electricity.

Tips for building owners

- Selecting an air-conditioning unit of the right size is important to optimizing energy consumption. As a rule of thumb, a 1-tonne unit is adequate for a room of 150 square feet (approximately 14 m^2), and a 2-tonne unit, for a room of 300 square feet (approximately 28 m^2).
- Outdoor units of split or window air-conditioning systems should be located in shaded areas with good air flow to make the system more efficient. Also, the outdoor units should be away from any source of heat source such as a chimney or a diesel-operated generator.
- Windows and doors of air-conditioned buildings should be designed so that they are sealed when shut to prevent air from leaking or entering closed spaces.
- As mentioned earlier, avoid setting the thermostat too low: choose a setting between 27 °C and 30 °C, taking into account

the preference of the occupants. This one measure alone leads to substantial savings.

- Regular maintenance helps in improving the energy efficiency of the systems.
- In summers, switch on the air-conditioning system in the early hours, before the building gets heated.

Integrating renewable energy with building design

Now that we have covered climate-responsive building design and energy efficiency, the final step to make a building truly 'green', with minimum impact on the environment, is to enlist renewable energy.

Depending upon the size of a project, some proportion of its electricity demand can be met through renewable energy. If enough space is available and if energy demand is optimized, even 'net zero' projects are possible: 'net' zero either because they are totally self-sufficient in terms of energy or because even if they draw some electricity from the grid, they generate a comparable amount of electricity through renewable energy sources.

The three main sources of renewable energy that can be integrated with buildings are solar energy, wind energy, and bio-energy (energy generated from organic waste generated within the building itself).

Solar energy

Solar energy systems that can be integrated with buildings are either solar PV (photovoltaic) systems or solar thermal systems (Figure 34).

Solar photovoltaic systems

Solar photovoltaic systems have been particularly successful in India because India receives solar radiation of high intensity: these systems convert incident solar radiation into electricity using semiconductors. The components of a solar PV system depend upon its configuration.

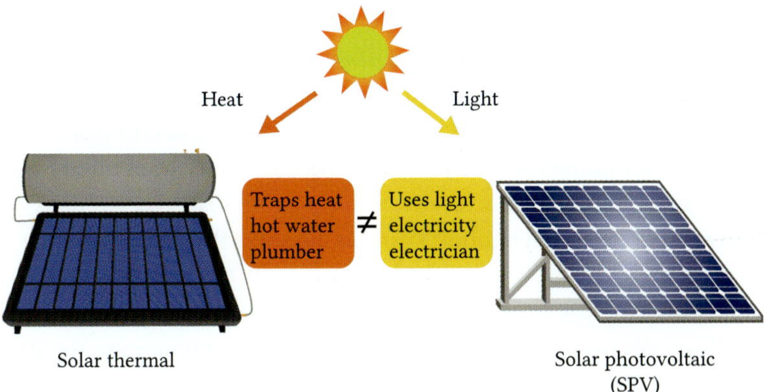

Figure 34 Solar photovoltaic and solar thermal systems

Solar thermal: Absorption of sunlight and conversion to heat. Higher ambient temperature enhances performance.

Solar PV: Semiconductor device, converts directly the incident photon energy into electricity. Higher ambient temperature deteriorates performance.

• *Grid-interactive solar photovoltaic systems* Government policies have made grid-interactive solar PV systems particularly popular in India. These systems are connected to the electricity grid, and electricity generated by the PV system as DC (direct current) is converted into AC (alternating current) at the grid voltage through a specially designed inverter. Such systems may or may not be coupled with batteries to store excess electricity. For more information on these systems, refer to *Sun Through the Roof,* which is part of the 'Concerned Citizen' series, as is the present book.

• *Building-integrated photovoltaic panels* Solar PV systems that are integrated with a building's structural system are known as BiPV (building-integrated photovoltaic) panels. In new buildings, BiPV panels are integral to the design and construction phases of the buildings, although the panels can also be integrated through retrofitting of existing buildings.

Photovoltaic systems can be used in many ways. For example, PV 'trees' can form part of a landscape, integrated into bus shelters and facades of buildings, with rooftops, as solar shades, as a roof system for atriums instead of glazed glass, and so on.

Integrating solar PV with buildings has many advantages.

- Generating electricity at no additional land cost. Current policies also allow building owners to sell excess electricity generated on site to utility companies.
- Reducing the load on cooling systems, as solar PV panels also act as shades.
- Achieving savings in materials, as BiPV will replace such construction material as glazing systems or some components of the roof, depending upon the design.
- Reducing the load on lighting to some extent, as BiPV panels allow some daylight to pass through them.

Tips for building owners

Solar PV systems should be designed based on the climate, pattern of solar radiation, and the building's power requirement. Install the panels at a spot free of any shadows. On flat roofs, the panels are mounted at an angle, which varies with the latitude of the location, so that they are exposed to maximum incident solar radiation throughout the year. For maximum exposure, the angle should be equal to the latitude of the location. In the northern hemisphere, the panels face south.

The construction should be strong enough to take the load of the PV panels.

The panels need to be cleaned as required, because dirt and dust on the surface of panels lower their capacity to absorb incident radiation.

Subsidies may be available through such schemes as SRISTI (sustainable rooftop implementation for solar transfiguration of India) by the central government, which seeks to promote solar rooftop technology among residents.

Solar water heaters
Solar water heaters use heat from the sun to provide running hot
water for various end uses. The system comprises solar collectors
to absorb incident solar radiation, insulated tanks to store hot
water, plumbing to take in cold water and to deliver hot water, and
other controls and instrumentation as required.

A solar water heater with a capacity of 100 litres per day is
sufficient for a family of four to five and can easily replace a 2 kW
electric geyser. Replacing an electric geyser with a solar hot water
system may easily save about 1500 units of electricity per family,
providing a quick payback within 2–3 years, depending upon the
charges for electricity and the quantity and the temperature of hot
water used. Solar water heaters can easily last for 15–20 years.

Technologies for solar collectors

Solar water heaters can be stationary, or fixed, or can swivel to
follow the sun, tracking it as it moves across the sky to exploit the
radiation fully.

Stationary solar collectors can be either in the form of flat
plates or comprise interconnected glass tubes. Flat-plate collectors
(Figure 35) are typically used for hot water below 80 °C; their
efficiency varies from 60% to 70% and their life, from 15 to 20 years.

Figure **35** Flat plate collectors

Evacuated tube collectors (Figure 36) comprise glass tubes filled with vacuum to minimize heat losses through conduction, which makes them not only more efficient than flat-plate collectors but also capable of delivering water at higher temperatures.

The highest temperatures are achieved in parabolic trough collectors (Figure 37), which comprise a series of concave mirrors that not only collect solar radiation but focus it, or concentrate it, on a narrow space, typically a receiver tube filled with water. The

Figure **36** Evacuated tube collectors

Figure **37** Parabolic troughs

series of mirrors is on the north–south axis to maximize incident radiation. Parabolic trough collectors can attain temperatures up to 400–600 °C and are used mostly in industrial processes and similar large-scale uses, including steam-based cooking on institutional scale.

Tips for building owners
Solar water heaters are generally installed on roofs or terraces; however, if space is not available, they can also be installed on south-facing facades.

Solar hot water systems should be installed in shadow-free spots.

For maximum energy collection over the year, the tilt of the solar collector plate should be equal to that of the latitude of the location.

A 100-litre capacity single collector will be about 2 m² and a footprint of about 3 m².

Depending on its capacity, a solar water heater full of water weighs approximately 1.2–1.4 kg per litre of water; therefore, the system should be strongly anchored in place.

The set point of the storage-type electrical water heaters should be between 50 °C and 60 °C.

Although solar water heaters require little maintenance, they need a continuous supply of water for smooth functioning. If they are not required for long periods, they should be covered with a non-transparent material to avoid overheating of collectors.

Solar water heaters may need a back-up system for cloudy days or when hot water is required in larger quantities.

Plumbing requires careful attention to minimize losses. If pipers are inadequately insulated or the water has to cover long distances, heat losses can make the operation less efficient.

Use of hot water also needs to be regulated: as hot water is drawn, cold water flows in to take its place, and drawing hot water in the evening can result in only lukewarm water the next morning.

Wind energy

Wind turbines work by converting the kinetic energy of wind into torque (turning force) acting perpendicular to the wind direction by using aerodynamic blades. The mechanical energy is then converted into electrical energy and, after further transformation, is matched to the frequency of supply grid and fed into it.

Wind turbines fall into two types, horizontal axis or vertical axis, depending on whether the axis around which the blades rotate is parallel to the ground or at right angles to it.

Horizontal-axis wind turbines are more common for large-scale operations. These are typically located in windy and open areas and feed electricity into the grid.

Vertical-axis wind turbines are smaller and typically more suitable for integrating them with buildings: they are quieter, less intrusive aesthetically, and can generate power even at low wind speeds, starting at about 3 metres per second (approximately 11 km/h). The power that a wind turbine can extract from wind depends on the cube of wind speed. However, in urban areas, the presence of buildings and other infrastructure limits the speed of wind, and it is advisable to study whether it is feasible to harness wind energy at a given site.

Building-integrated wind turbines typically depend on the building itself for support – whereas large-scale, horizontal wind turbines are supported by towers specially erected for the purpose – which is why the building needs to be designed to take the additional load. Hybrid systems are also available, which combine wind turbines and solar PV systems (Figure 38).

High-rise buildings have good potential to integrate wind turbines into the building design because wind speed increases with height (Figure 39).

Tips for building owners

Buildings that seek to harness wind energy should be energy efficient in the first place.

Figure **38** A hybrid system combining vertical-axis wind turbines with solar photovoltaic panels

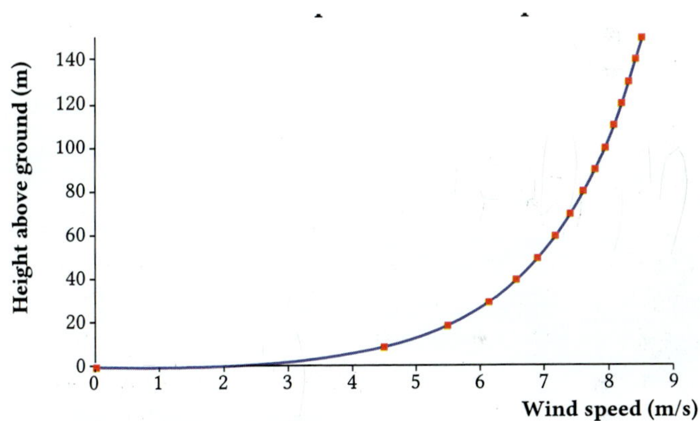

Figure **39** Vertical profile of wind speed (s o u r c e https://wingardium-energy.com)

The overall impact of deploying wind energy on the local environment or the neighbourhood needs to be assessed. For example, wind turbines can be noisy: those that are to be located right in the middle of a city need to be chosen for their quiet operation.

Bio-energy

Occupants of buildings can generate a great deal of organic waste, which can be a source of energy. Biomass gasification is a renewable energy technology because it converts such locally available bio-resources as food waste, wet waste from municipalities, crop residues, and dried twigs, branches, and leaves from forests into combustible gas that burns cleanly. Biogas is a mixture of gases, mainly methane and carbon dioxide, produced by the breakdown of organic matter in absence of oxygen (anaerobically).

Biogas or biomethane has a calorific value higher than that of LPG (liquefied petroleum gas) and is used as a fuel for cooking, transport, and power generation.

Small biogas plants (Figure 40) are appropriate for buildings and convert organic wet waste into biogas, which can replace LPG for cooking. Such plants have a small footprint, as small as 20 feet by 40 feet (approximately 6 metres by 12 metres). Each tonne

Figure **40** Installations by CarbonLites to convert waste into resource

of waste generates gas equivalent to about 50 kg of LPG and 100 kg of manure as a by-product every day. Such biogas plants have found applications in buildings such as hotels and restaurants, apartment blocks, temples, information-technology parks, and universities.

The various components of green buildings are summarized in Figure 41.

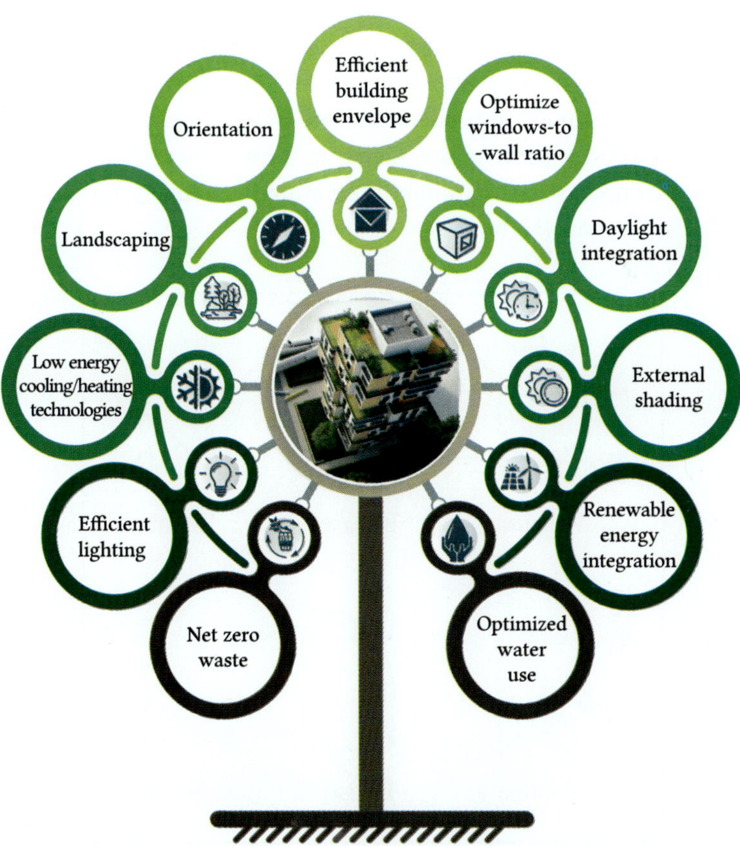

Figure **41** Features that can be incorporated in green buildings

Waste

Managing waste is like creating a problem and then trying to solve it. Before buying anything, be it an item of clothing, food, a piece of furniture, or a fancy electronic item, how often do we ask ourselves: Do I really need it? Can I make do instead with refurbished clothes or a bit of alteration here and there? Can we stop throwing away foodstuff and instead finish whatever is on our plate? Can we avoid buying that extra piece of furniture or, for that matter, not switch to a fancier model of a mobile phone? The world is facing the immense challenge of managing the waste that we create from our day-to-day activities, be it food waste that goes into landfills, packaging waste such as single-use plastics, or construction and demolition waste. It is high time we started moving from a linear approach to waste management – in which we only consider waste management in terms of recycling, reuse, and reducing waste – to adding three more R's, namely rethink, refuse, and repair (Figure 42).

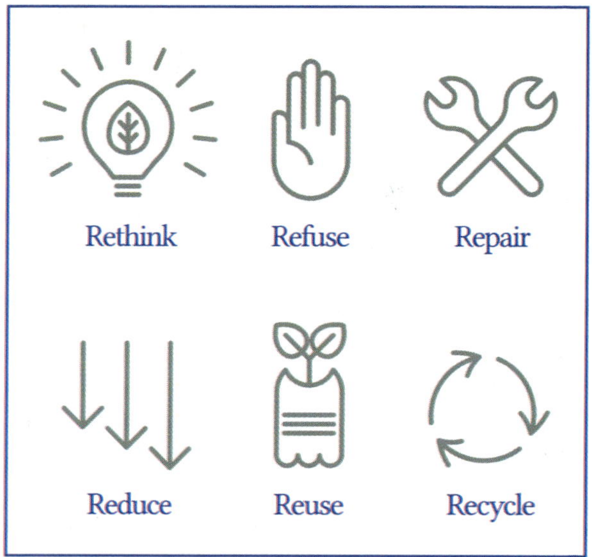

Figure **42** The six R's of sustainability

Within the hierarchy of waste management, the six agreed-upon principles of sustainability are rethink, refuse, reduce, reuse, repair and recycle.

Rethink

Do I really need that? One should always ask oneself if a purchase is justified based on one's consumption needs and the likely impact of that purchase on the environment. If we realize that resources are limited, we are likely to rethink our everyday choices.

Refuse

To reduce the amount of waste we generate, we should try to refrain from buying things we do not need, and buy products based on certain principles that help in reducing waste. For example, we can buy products based on their quality, the packaging, and the ethics of the company that makes them. Essentially, we should refuse to buy things that add to the waste. For example, we should avoid buying a small item if it is packed using a lot of plastic and paper. Refusal helps to prevent materials and products from entering homes and offices in the first place. Let us see how we can deploy the 'refuse' principle in our day-to-day life, be it at home or in the workplace.

Refuse unwanted free gifts or printed literature distributed at conferences, trade shows, etc.

Refuse single-use plastics or disposables such as paper cups, razor blades, stirrers, and bottled water and opt for reusable materials instead.

Reduce

If we cannot refuse, we should at least try to reduce consumption and thus partially reduce waste generation. For the uninitiated, it means reducing the amount of stuff you need and actively making decisions that reduce waste. In other words, consume less, waste less. Let us look at a few methods to reduce waste in the workplace and at home.

- Free seating Reduce the number of workstations by deploying free seating arrangements.
- Double-sided printing Print on both sides of a sheet and also avoid printing if the matter can be read off a screen.
- Used paper Retain the sheets printed only on one side and, when they are no longer required, use them for printing fresh batches of text.
- Digital documents Store documents digitally rather than using paper-based storage, which not only consumes a lot of paper but also takes up valuable physical space. Similarly, we can use digital contracts in place of paper-based contracts to reduce the use of paper.
- Reusable utensils Use plates, glasses, cutlery, etc. that can be used again and again to reduce waste generation.
- Shopping bags Carry your own bags when you shop for groceries, vegetables, etc. and decline plastic carry bags offered by vendors. Try to buy material in bulk and use refillable containers to store detergents, perishables, beverages, etc.
- Biodegradable products Switch to using cleaning agents that contain less volatile or non-volatile organic compounds, which are gentler on the environment. Look for concentrated solutions of cleaning agents to reduce your waste even more: most of them are required only in small amounts, to be diluted with water before use.
- Responsible transport Walk or cycle to work, organize or join car pools, and use teleconferencing whenever possible.
- Local products From coffee and office furniture to cleaning supplies and other office supplies, use a local supplier to limit the distance your goods must travel.

Reuse

If you don't need something and you are thinking of throwing it away, try and see if you can reuse or repurpose it. Instead of buying a replacement, find alternative uses for the stuff to be discarded.

o Go for reusable water cups or bottles instead of plastic bottles.
o Refill your printer cartridges instead of buying new ones.
o Reuse boxes and packaging material from the courier and e-commerce sites.
o Purchase cloth towels for the office kitchen instead of paper towels.
o Repair broken or faulty office equipment or other items.
o Donate to charitable organizations for use by the needy.

Repair

Try to repair and fix electronic gadgets, furniture, watches, mobile phones, etc. before deciding to buy new stuff.

Recycle

If you really cannot reuse something, recycle it. You have no excuse not to recycle everything you can in the present times. By separating your waste, you help in ensuring that it reaches the right treatment centres. The raw materials in such products can be reclaimed and reused to make other products, which means you're not using any new reserves of natural resources and contributing to sustainable development.

Waste-treatment technologies

Thermal treatment of waste makes it possible to turn it into useful products such as compost and energy. Waste treatment techniques are not dealt with in the book and can be explored by the users.

Tips for building owners

- Divert wet and organic waste from reaching landfills; instead, treat such waste to produce compost (the aerobic process) or biogas (the anaerobic process).
- Generate less waste.
- Have segregated waste collected more efficiently.
- Avoid the release of methane into the atmosphere by treating wet waste aerobically to produce compost or anaerobically to produce biogas.
- For large-scale processing of wet waste, use commercially available biodigesters; these can handle 2–20 kg of organic solid waste a day.
- For establishments that generate waste in bulk, such as hotels, apartments, temples, IT parks, restaurants, and schools and colleges, install biogas plants of suitable size. Each tonne of wet waste generates biogas equivalent to about 50 kg of LPG and gives 100 kg of manure as by-product every day. A small biogas plant can be set up over an area measuring roughly 6 metres by 12 metres (source CarbonLites).
- Use landfills as the last resort, to be used for what is left over after the waste has been treated.

Conclusion

The value of being 'green' is best realized early on. Green buildings are designed to use energy and water optimally, generate minimal waste, and provide a productive and healthier work environment. They provide air of higher quality, they impact the land minimally, and, with proper planning, they are in every way the economically smarter way to go.

Whether it is the integration of passive design principles at the preliminary design stage including building site orientation, daylighting, and natural ventilation or the selection of building materials and finishings with low-embodied carbon that reduce and eliminate dangerous volatile organic compounds from the air we breathe or integrated system design and optimization so that a building conserves precious natural resources such as energy and water, green building strategies have been transforming how we think about buildings and the spaces around us for decades.

Green buildings come with various levels of sophistication, but all citizens can do their bit: it is the collective effort that matters. Although best integrated at the design stage, even retrofit options and efficient lifestyles can help save resources. This book is an attempt to demystify the concepts and empower everyone to think and live green.

As green building practitioners on the frontline of market transformation in India, we believe that our homes, buildings, and communities must move from not just doing less harm to becoming truly regenerative.